INTRODUCTION TO
PROFESSIONAL FOODSERVICE

INTRODUCTION TO PROFESSIONAL FOODSERVICE

Wallace L. Rande, Ed.D.
School of Hotel and Restaurant Management
Northern Arizona University

JOHN WILEY & SONS, INC.
New York • Chichester • Brisbane • Toronto • Singapore

Chapter Opening Photo Credits

2, p. 26, Courtesy DAKA, International, Inc.; 3, p.46, Courtesy Gamma Liaison; 4, p. 64, Courtesy National Dairy Council; 5, p. 82, Stephen Graham, Educational Foundation of the National Restaurant Association; 6, p. 106, Courtesy W.R. Grace & Company; 7, p. 130, Courtesy Fudderuckers, Inc.; 8, p. 154, Courtesy Hobart Corporation; 9, p. 186, Farrell Grehan Photo Researchers; 10, p. 200, Photo Jim Holland/NYT Pictures; 11, p. 222, Courtesy Jack D. Mezrah; 12, p. 240 Courtesy Marriott Food & Services Management

This text is printed on acid-free paper.

Library of Congress Cataloging in Publication Data:
Rande, Wallace L.
   Introduction to professional foodservice / Wallace L. Rande.
     p.   cm.
   Includes bibliographical references.
   ISBN 0–471–57746–4 (cloth : alk. paper)
   1. Food service. I. Title.
TX911.R33 1995   95–13388
647.95--dc20

Printed in the United States of America

10 9 8 7 6 5 4 3 2 1

CONTENTS

PREFACE

This book is intended as a text for introductory courses in foodservice management. Its chapters provide, first, a brief history and an overview of the foodservice industry. Then they cover the fundamental areas of a foodservice operation—menu planning, purchasing and receiving, production, services, safety and cleaning, and cost management.

The book also contains several chapters on topics not normally emphasized in foodservice management texts: the role of the customer, nutritional concerns, and organization theory and systems. The foodservice business is very dynamic and always changing. These chapters were included to reflect trends that, I believe, will continue to be crucial in our industry.

To succeed in their increasingly more competitive arena, foodservice operations need to be customer-driven. Throughout, this text stresses the importance of the customer to the success of a foodservice operation. Chapter Two is dedicated specifically to the role of the customer. The customer, surprisingly, has been left out in most foodservice textbooks, but future managers (and those currently working in the industry) must respond to their customers in all aspects of the running of their business. This customer service orientation is reinforced at the beginning of each chapter by relating the importance of its topic to the guest. This is because I believe that the student's foodservice education is greatly enriched by an awareness of the role and importance of the customer.

Chapter 3, on foodservice organization and systems, provides a brief history of the evolution of management and organizational theory. Various foodservice systems are used to illustrate the different paths that food travels from preparation to the server to the guest. It is interesting to see how the problems encountered in different foodservice operations are handled within these foodservice systems.

Concern about nutrition was once considered a popular fad that would soon fade. Then it was frequently argued that foodservice operations need not be concerned with nutrition and with providing healthier meals; guests eat only one out of four meals out and can "eat healthy" at home. The trend toward nutritional awareness has endured, however, and customers routinely expect lower-fat, lower-calorie, and more nutritious choices on foodservice menus. Chapter 4, on nutrition and the foodservice industry, orients the topic of nutrition to students and demonstrates to them the different views held by guests toward nutrition when they are dining out.

This text is organized in a manner that facilitates both teaching and learning. Chapter objectives are presented at the beginning of each chapter, to prepare the reader for what follows and to identify the key concepts. The chapters conclude with lists of key terms, discussion and review questions, suggested readings, and, periodically, activities to reinforce the material. A comprehensive glossary is included at the end of the book. Main points are highlighted, defined, and explained in the text.

I am grateful to the educators who reviewed both the proposal for this book and early drafts of the manuscript, and to my students, for all that I have learned from them.

Wallace L. Rande, Ed.D.

INTRODUCTION TO
PROFESSIONAL FOODSERVICE

CHAPTER ONE

THE FOODSERVICE INDUSTRY

CHAPTER OBJECTIVES (Questions)

After reading this chapter you should be able to:

- Discuss the historical roots of the foodservice industry.
- Describe the three forms of foodservice ownership.
- List the types of foodservice operations that do business in the commercial segment of the industry.
- State the importance of foodservice operations in lodging facilities.
- Discuss the growth of foodservice in grocery and convenience stores.
- List the types of foodservice operations that do business in the noncommercial segment.
- Describe demographic trends and tell how they affect the foodservice industry.

To understand and appreciate the foodservice industry, students must be acquainted with its history, scope, and trends. A prospective manager should know where the industry has been and what its roots are, be aware of the sizes and types of operations that make up the various segments of the industry, and, finally, gain an idea of where the industry is going by examining trends and recent developments. The purpose of this chapter is to acquaint the student with the range of variation in the industry, as well as the similarities and differences between the various types of foodservice operations.

Customers have always played a key role in the foodservice industry. The first foodservice evolved to meet the needs of an increasing traveling population. New foodservice operations and new segments of the industry grew, and continue to grow, in response to the demands of travelers. For example, in-flight catering services were started when a certain airline discovered that flight passengers and crews were stopping at a Marriott restaurant before boarding a plane in Washington, D.C. The airline decided that this was a service it should offer, so it contracted with Marriott to provide food for all its passengers.

HISTORY OF THE FOODSERVICE INDUSTRY

Quantity foodservice as we know it today is a fairly recent occurrence in the history of the world. When the economic base of the world was agricultural, people's livelihood depended on farming; they lived mostly off the land and had little time or need to travel. With the beginning of the Crusades and the flourishing of commerce, travelers needed places to sleep and eat while away from home. Monasteries and other religious establishments opened their doors to wayfarers. Inns and taverns, spaced a day's journey apart, began to appear along trade routes to provide the bare necessities for weary travelers.

Europeans brought their system of inns and taverns with them when they settled America. The first tavern in America opened in 1634. In 1656 the hospitality industry was given a boost when the colony of Massachusetts required all towns in the colony to have a tavern. Taverns popped up along the trade routes and flourished as the population increased and people began to travel more widely. The role of the tavern in society then began to shift from providing a place away from home for travelers to being a social gathering place for the town. Taverns became the center of social activity and formed a vital communication network for the colonies. The American Revolution was planned in these taverns, from which was born the American hospitality industry.

The inns of the growing country instituted what was called the **American plan**, in which the price of a room included meals. The food served in these early inns was mostly European style, as Americans were not known for their culinary expertise. However, many guests, especially European guests, did not like the idea of having to pay for meals whether they ate them or not. The Parker House in Boston was the first inn to break tradition and offer the **European plan**, in which rooms and meals were charged separately.

As the towns of early America became more self-sufficient,

people traveled less and the village tavern gave way to the next trend of the foodservice industry, the restaurant. In 1820 the first three restaurants of the nation opened in New York City: Delmonico's, Niblo's Garden, and Sans Souci. Delmonico's emerged as an American tradition and is generally regarded as the first American stand-alone restaurant.

The 1800s represent an important era in the development of commercial foodservice in America. A few of the innovations of the time are shown in Table 1.1.

The primary mode of long distance travel in the first half of the nineteenth century was the train. Before the development of the dining car, trains would stop at stations along the route so that passengers could get off to eat. George Pullman, inventor of the railroad sleeping car, also developed the first dining car. Dining on the train was a luxurious affair; passengers were served the gastronomic delights of the era. This mode of dining became so popular that creative business people began to create stationary restaurants that looked like dining cars, located away from the railroad, which later developed into diner-type restaurants.

American John Krueger is credited with the development of the cafeteria-style restaurant. He borrowed the concept from the smor-

 TABLE 1.1 Foodservice Innovations of the 1800s

1803	The first ice refrigerator.
1815	The first industrial foodservice operation in our nation was begun when Robert Owen opened a dining facility for his workers and their families.
1825	The first gas stove.
1834	Bowery Savings Bank of New York began an employee dining program, which is still in operation today.
1855	Harvey Parker offered the first à la carte menu as part of the European plan at the original Parker House in Boston.
1860s	The development of the first dishwashing machine.
1860s	The first martini was made in San Francisco's Occidental Hotel.

Source: Adapted from Donald Lundberg, *The Hotel and Restaurant Business,* 5th ed. (New York: Van Nostrand Reinhold, 1989).

gasbord of his native Sweden. Cafeterias were first used during the Gold Rush days in California to feed miners, who wanted to eat quickly so that they could get back to work. On Wall Street, a cafeteria was opened to provide quick meals to workers in the growing financial center of the nation.

SCOPE OF THE FOODSERVICE INDUSTRY *answer for Canada*

CONTRIBUTION TO THE NATIONAL ECONOMY (4a)

The foodservice industry is an essential part of the economy of the United States. Total sales for the industry were forecast to be almost $289.7 billion in 1993. Food and beverage purchases for 1993 were expected to exceed $109 billion. Sales were expected to increase about 4.7% when adjusted for inflation 2.4% in 1994–1995 (see Table 1.2). The industry accounts for about 4.3% of the GNP. One out of four retail establishments in America are eating and drinking operations. The foodservice industry accounts for about 43% of the amount of money spent on food for the nation. Sales for foodservice have grown consistently in the past several decades (see Figure 1.1).

EMPLOYMENT (4b) *How many in Canada work Hospitality industry?*

Nationwide in the United States, foodservice operations employ approximately 9 million people, or 5% of Americans over 16 years old. Total foodservice employment is expected to reach 12.4 million by the year 2005. Food and beverage operations contribute more than $49 billion to the national economy through wages and benefits. Moreover, the foodservice industry provides opportunities for women and minorities: more than 66% of foodservice supervisors are women, and 20% of supervisory personnel are either black or hispanic.

TYPES OF OWNERSHIP

Eating and drinking establishments are mostly small businesses. About 72% of these operations have annual sales under a half million dollars. The average sales of restaurants and lunchrooms was about $428,000 in 1987, and the average for fast food restaurants was almost $412,000 in the same year. Three types of ownership arrangements occur in foodservice operations: franchises, chains, and independent ownership.

TABLE 1.2 Foodservice Industry Sales Forecast for 1995

Commercial Foodservice	Sales (in billions)
Eating Places	
Restaurants	$87.80
Limited menu/fast food	93.35
Commercial cafeterias	4.86
Social caterers	3.05
Ice cream/frozen yogurt	2.63
Total eating places	$191.69
Bars and taverns	9.35
Food Contractors	
Industrial sites	4.46
Office buildings	1.45
Health care sites	1.95
Colleges, universities	3.65
Schools	1.59
Airlines	1.60
Recreational sites	2.41
Total food contractors	$17.11
Lodging eating places	16.93
In-store restaurants	12.32
Recreation, sport sites	3.41
Mobile caters	.96
Vending, nonstore retail	6.22
Total Commercial	$258.00
Institutional Market	
Employee dining	1.79
Schools	4.39
Colleges, universities	5.00
Transportation sites	1.55
Hospitals	9.92
Nursing homes	4.42
Clubs, sport camps	2.41
Community centers	1.12
Total institutional	$30.57
Military	1.10
Grand Total	$289.68

Source: National Restaurant Association.

Statistics canada.com

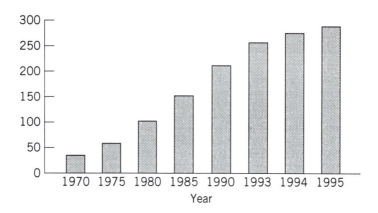

FIGURE 1.1 Foodservice industry food and beverage sales, 1970–1993. (*Source:* National Restaurant Association)

Franchises — owner owns it. Sheraton/Fairmont

privately owned

A **franchise** is an arrangement between two parties: the franchisor and the franchisee. The franchisor grants the franchisee rights to sell certain goods and services, to use various logos and promotional items, and to participate in national media campaigns. It is usual for the franchisee also to purchase operational systems and controls, as well as facility designs and layouts, with the franchise package. The franchisor must agree to maintain corporate standards and image in order to preserve the business agreement.

Advantages to franchisees for engaging in this type of arrangement are many. They are able to purchase the operating expertise and successful management systems of the parent company. Facility design, equipment layout, menu design, and recipe development are included in the package. The ability to use a nationally or regionally recognized name is also a plus, especially for operators who are new to the restaurant business.

Initial setup costs and costs of operation, however, are higher for a franchise than for other types of foodservice establishments. The franchisee must pay initial franchise fees and ongoing royalty fees for the duration of the business. Some franchise agreements require that businesses purchase goods from the parent company or its subsidiary for above-market prices. The goal of the parent company is consistency among all franchises; therefore, creativity and innovation in menu offerings are usually restricted.

Table 1.3 lists the top 10 foodservice companies in the United

Food service companies in Canada are?

TABLE 1.3 The Top 10 Foodservice Companies Ranked by Total Number of Stores

1992 Rank	1991 Rank	Chain	Current Total	Company Owned	Franchises
1	1	McDonald's	9,051	1,495	7,556
2	2	Pizza Hut	7,280	4,230	3,050
3	4	Subway	7,000	0	7,000
4	3	7-Eleven	5,789	2,759	3,030
5	5	Burger King	5,675	850	4,825
6	6	KFC	5,116	1,900	3,216
7	7	Domino's Pizza	4,860	1,050	3,220
8	8	Dairy Queen	4,665	5	4,660
9	11	Little Caesar's Pizza	4,575	1,525	3,050
10	12	Taco Bell	3,908	2,338	1,570

Source: Nation's Restaurant News Research, 1992.

States based on the number of stores they have in operation. Notice that the major chains own some of their units and that some of their units are operated by others through franchising.

Restaurant Chains *multi-united group = publically owned*

Another form of ownership is the **restaurant chain**, each operation being part of a multiunit group. The restaurant concept, design, menu, decor, and style of service are developed by the parent company and duplicated throughout the country. Chain restaurants are found more often in fast food or limited-menu operations rather than in full-service restaurants. It is simpler to standardize and reproduce restaurants with limited service and menus than those that offer full service. Chain restaurants may be owned by a parent company, a franchise company, or independent owners.

An advantage of operating chain restaurants is that some costs are consolidated. Once an efficient design is developed, it is duplicated in other restaurants in the chain, thus reducing the cost of facility development for each operation. Purchasing is usually pooled, again reducing costs. Advertising and promotions are coordinated to maximize their effectiveness. Menu development and printing are also consolidated to further reduce costs and maintain consistency.

A disadvantage of running a chain restaurant is that most

TABLE 1.4   The Top 10 Foodservice Companies Ranked by U.S. Sales

1992 Rank	1991 Rank	Chain	Parent Company	1992 U.S. Sales ($ millions)
1	1	McDonald's	McDonald's Corp.	12,703
2	2	Burger King	Grand Metropolitan PLC	6,450
3	4	Pizza Hut	PepsiCo Inc.	4,450
4	3	Hardee's	Imasco Ltd.	3,300
5	5	KFC	PepsiCo Inc.	3,600
6	7	Taco Bell	PepsiCo Inc.	3,100
7	6	Wendy's	Wendy's International Inc.	3,001
8	9	Marriott Management Services	Marriott Corp.	2,500
9	8	Domino's Pizza	Domino's Pizza Corp.	2,358
10	12	Little Caesar's Pizza	Little Caesar's Pizza Inc.	2,160

Source: *Nation's Restaurant News* Research, 1992.

management decisions are made at the top corporate level, far from the locations of many of the restaurants. The diversity of the various regions of our nation generally makes standardization of menus and food items difficult. For example, what sells well as "Southwestern" food in Kansas City may not be as successful in Texas. In addition, the corporate structure may pose difficulty making decisions on a timely basis so that chains can keep up with an ever-changing market place. Table 1.4 lists the top 10 foodservice companies based on sales in the United States.

Independent Ownership *privately owned*

A third type of ownership is **independent ownership**. Three of every four eating and drinking establishments are single-unit operations, and about one-third are **sole proprietorships**; that is, they are owned by one person. Although chains and franchises are continuing to expand in the industry, there will always be room for the hard-working and creative **entrepreneur** to own and operate his or her own business.

The down side of independent ownership is that the majority of such businesses fail within their first few years of operation. The chance of remaining in business for 3 years is 50%, and for 10 years only about 30%. Several factors contribute to this high rate of failure, including lack of adequate funding and insufficient knowl-

TABLE 1.5 The Top 10 Independent Restaurants in the U.S. Ranked by Sales

	Name/City	1992 Total Sales	Average Dinner Check
1	Tavern on the Green, New York, NY	$26,959,000	$44.00
2	Bob Chinn's Crabhouse, Wheeling, IL	$16,314,382	$124.90
3	Phillips Harborplace, Ocean City, MD	$13,664,576	$20.85
4	Frankenmuth Barvarian Inn, Frankenmuth, MI	$12,356,861	$11.50
5	Zehnder's, Frankenmuth, MI	$12,265,000	$12.50
6	Joe's Stone Crab, Miami Beach, FL	$11,111,000	$35.00
7	Spenger's, Berkeley, CA	$11,100,000	$12.00
8	Shooters, Fort Lauderdale, FL	$10,769,000	$17.25
9	Montgomery Inn at the Boathouse, Cincinnati, OH	$10,100,000	$16.50
10	Rascal House, Miami Beach, FL	$8,522,590	$8.98

Source: *Restaurant Hospitality* Research, 1992.

edge of the foodservice industry. Some people mistakenly believe that if they have eaten in restaurants, they will be able to operate one, or that they can serve what they like without considering customer demand. For example, if a person buys a pizza restaurant but serves only his or her own favorite type of pizza, chances are that the business will close within the first year.

Independent restaurants have an advantage over chain and franchise restaurants in that they are closer to the people they serve and so can react faster and more efficiently to changes in the local market. However, they may be at a disadvantage as compared with chain and franchise restaurants because they do not have corporate support. The top 10 independent restaurants are listed in Table 1.5.

CLASSIFICATIONS OF THE FOODSERVICE INDUSTRY

To better understand the foodservice industry in America, it is useful to examine the categories and classifications used by the **National Restaurant Association (NRA)**. The categories discussed here are the commercial and the noncommercial operations (see Figure 1.2).

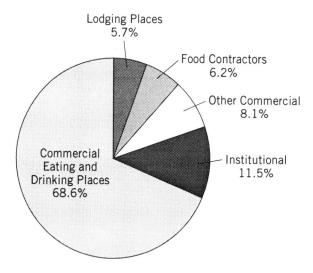

FIGURE 1.2 Foodservice industry segments by percentage of sales, 1993.

COMMERCIAL OPERATIONS

The largest segment of the foodservice industry is the **commercial, for-profit**, group. Operations in this category are designed to earn revenues in excess of expenses so as to provide a financial return to investors and owners. Operational decisions are often based on how they affect the **bottom line**. Key segments of the commercial category are eating and drinking establishments: restaurants, commercial cafeterias, social caterers, and ice cream and yogurt stores (see Figure 1.3).

Restaurants range from full-service, sit-down restaurants to walk-up, self-service, limited-menu establishments. These operations differ in their type of service, type of food served, decor, and prices.

Full-service restaurants provide table service and usually feature a variety of items on their menus. They can be either independently owned or part of a national chain or franchise. The majority of full-service restaurants are independently owned and operated. Businesses in this category share some common features. They generally produce most of their food from raw ingredients on the premises and require relatively highly skilled preparation and service personnel. The **average check** amount ranges from moderate to high. Meals are served in multiple courses; the service is

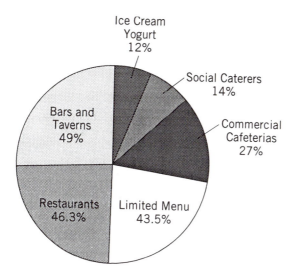

FIGURE 1.3 Commercial foodservice: eating places by percentage of sales, 1993.

relaxed and unrushed. Generally, this type of establishment does not need to rely on a high customer turnover rate to maintain an adequate level of profit. Alcohol service is usually provided to complement the menu and to provide additional revenue. The service staff is a very important component of this type of restaurant because of its impact on the customers' enjoyment of their dining experience. Customers patronizing a full-service restaurant expect more than just food to fill their stomachs; they are seeking a pleasurable experience along with their meal.

Fast-food (or **quick service**) **restaurants** have existed in different versions for quite some time. Roadside diners, hamburger stands, and Automats dominated American foodservice of yester-year. As society became more mobile through the increased availability of automobiles, along with the development of roads and parking lots, the fast-food segment, spurred on by the introduction of franchising, began to grow.

Fast-food restaurants nationwide share some common characteristics. They are able to provide food quickly owing to the reduced production requirements of the food they serve. Labor requirements are also reduced because customers either go to the counter to order food or "drive-thru" in their cars to order and receive their meals. The average check amount is generally low, with the result that the business must rely on substantial volumes of customers as well as high turnover rates of seats in the dining room in order to generate an adequate profit. The menu is limited to multiple variations of a few types of items, such as pizza, chicken, hamburgers, fish, or ethnic specialties. The items on the menu are usually priced separately, with some combination meals offered as a special value for customers. Fast-food operations use a high percentage of convenience foods to reduce the number of skilled preparation personnel and the amount of production equipment required in the kitchen. The kitchens are more highly automated than those in full-service operations, which also increases productivity, reduces labor needs, and keeps costs down. Fast-food operations use disposable utensils and dinnerware to eliminate or reduce the need for dishwashing facilities and to complement their takeout business.

Commercial cafeterias are set up similarly to school cafeterias. The food is pre-prepared and served from **steam tables** to customers, who walk through a line and pay for the total of individual items. Unlike school cafeterias, however, commercial cafeterias provide more menu choices and are open to the public. This type of operation is popular in the southern regions of the country. Customers like the ability to have control over the price and composition of their meals.

Social caterers prepare food in one location and then deliver and serve it at another location (such as the customer's home or office). Services provided range from simple food and beverage preparation and service to complete party planning and setup. Social caterers offer a convenient way for individuals and companies to entertain.

Ice cream/frozen yogurt stores are also considered as part of the "restaurant" category by the National Restaurant Association. The nation's passion for premium ice cream and frozen yogurt fuels growth in this area of the foodservice industry. Ice cream and frozen yogurt shops are opening all over the country in order to satisfy the demand for these products. It is noteworthy that the continued growth in the popularity of ultrarich, premium ice cream contradicts Americans' continued interest in reducing their fat consumption. Low-fat or nonfat frozen yogurt, on the other hand, provides consumers a healthy choice for a frozen dessert, as its increased popularity attests.

The **bar and tavern** segment of the foodservice industry is the only area that has experienced a decline in sales. Despite the popularity of sports bars, social pressure has resulted in a decrease in alcohol consumption. Creative operators are trying to supplement revenues by expanding food menus and nonalcoholic beverage offerings.

Food Contractors *Marriot*

Food management companies have experienced steady growth over the last few years. Increasingly, businesses such as company dining rooms, schools, hospitals, and sport centers are hiring outside contractors to provide foodservice within their facilities. The advantage is that the organization can concentrate on what it does best and eliminate the extra effort of managing a foodservice operation. The advantage to the **food contractor**/management company is that it is able to expand its business without a large outlay of capital. The cost to the management company is limited to staffing/management, operation costs, and possibly a few site improvements. Some food management companies are adding complementary services such as dry cleaning, convenience stores, and day care to better serve their customers.

Generally, a company seeking a contractor requests bids. The major national companies involved in this segment of the foodservice industry are Marriott Management Services, ARAMARK, Canteen, and Host International, as well as several strong regional companies.

Foodservice in Lodging Places

Foodservice plays an important role in lodging places. Whether a catering/banquet department, room service, coffee shop, or full-service restaurant, it offers a valuable service to the guests of a hotel. On the average, about 34% of total hotel revenues are provided by food and beverage operations, although this accounts for only about 18% of profits.

The food and beverage division of a hotel can be a valuable drawing card for local business, as well as being important to the operation of a hotel. Patrons of hotel food and beverage outlets are potential customers for the other departments and services the hotel has to offer.

Restaurants in a hotel serve both the guests and the local community. Eating at the hotel is convenient for guests, and local residents dine at hotel restaurants while visiting friends or business associates staying at the hotel. Hotel restaurants also serve as an alternative to other local restaurants. Buffets, brunches, and holiday meals allow the culinary and foodservice staff of the hotel to show off their expertise, and these events also provide a service to hotel guests and the local community.

Room service is an important amenity for hotel guests. Guests traveling alone, unfamiliar with the town, or just wanting privacy appreciate the convenience of having their meals served in their rooms. Luxury hotels in large metropolitan areas provide room service 24 hours a day to accommodate any traveler's schedule. Room service menus are adapted from the hotel restaurant's menus and priced at a premium to cover the increased labor and equipment required. The food is prepared either in restaurant kitchens or, if room service departments are very busy, in the hotel kitchen.

A **catering/banquet** department serves both hotel guests and the local community. Most banquet rooms are designed to accommodate a wide variety of group sizes, allowing maximum flexibility. For example, a hotel's grand ballroom may accommodate 500 people or more when fully open, or it may be sectioned into halves, quarters, or smaller rooms to serve parties of various sizes.

Banquets and catered functions are highly profitable and represent a large portion of revenue for the hotel food and beverage department. Catered functions held at a hotel help to generate room occupancy by offering a convenient place for guests to stay while attending a meeting or banquet. Weddings, banquets, and community events catered by the hotel help to bring members of the local community into the hotel and may lead to further business in room occupancy. The catering department also provides refreshments for business meetings held at the hotel.

Grocery and Convenience Store Foodservice

The fastest growing segment of the industry comprises **grocery store foodservice** and **convenience stores**. The growth of the foodservice business in both types of stores is propelled by changes in our society. Many new full-service grocery stores include fully stocked delicatessens, as well as soup, taco, and salad bars. Most stores provide places for customers to eat, or package food for takeout. Some stores are leasing space to fast food companies to broaden the food choices for their customers. An example of this type of development is Taco Bell's opening of small operations in a chain of grocery stores and gas station convenience stores in Phoenix, Arizona.

Dual-income families are patronizing takeout counters at local grocery stores as an alternative to eating in restaurants. They enjoy the privacy, convenience, and economy of eating at home, part of which is not having to pay a service charge or leave a tip.

Convenience stores have emerged as the "one-stop shopping" places of the 1990s. Most provide everything from gasoline to hardware to groceries and takeout food for people on the go. Customers pay a premium price for the expedience of service. These stores sell convenience foods both fresh and frozen, along with a variety of drinks, and often have a microwave oven for customers to heat up their meal choices.

NONCOMMERCIAL FOODSERVICE OPERATIONS

The goal of a **noncommercial** operation is, primarily, to generate enough revenue in sales to cover the expenses of the operation and provide money for replacement and repair of equipment and facilities, rather than to make a profit. Operators still have to pay very close attention to costs so that they stay within budget, especially with rising costs and funding cutbacks. Operations in the noncommercial segment are usually institutional in nature (hospitals, penal institutions, the military, schools, and universities), providing foodservice in-house rather than contracting with an outside foodservice provider (see Figure 1.4).

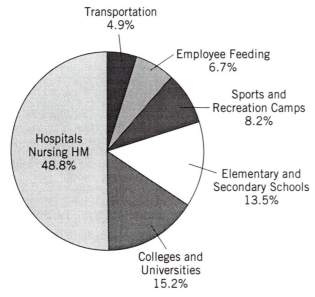

FIGURE 1.4 Institutional foodservice segments by percentage of sales, 1993.

College and University Foodservice

Colleges and universities have traditionally provided meals and rooms for students. Through the years, however, the type of meal service and the operation of college/university foodservice has changed. Currently, foodservice at most universities and colleges is provided either by a food contractor or management company, or internally by a department of the school.

Options available to students have changed dramatically over the years. Board plans in the past required students to purchase a full meal plan, with little flexibility. The foodservice operators had little competition from local restaurants because students paid for their meals at the beginning of the semester and did not receive a refund for uneaten meals. Student dining halls in the past offered a choice of only two main courses, each including a soup, a vegetable, and a starch. Today's school foodservice offers many more choices and may include soup, salad, deli, and dessert bars to appeal to a wider range of student tastes. In some situations the students are a **captive audience** for these foodservice operators. The responsibility in providing food to these students is much greater than in providing food to those who have other options. Special care must be taken to ensure that balanced meals are offered and that students can make healthful choices. The responsibility, however, is limited to just that, because it is the student who ultimately decides what his or her choice of meal will be.

There is also much greater flexibility in meal plans today. Students at most schools have a choice regarding the number of meals they would like to purchase. Most schools offer students options of either 0, 10, 15, or 20 meals per week. The variety of foods offered has also grown dramatically. Self-serve salad, sandwich, and soup bars have increased students' choices. The "healthy food" craze has also filtered into many college dining halls, which may now offer vegetarian specials and low-fat and low-cholesterol items.

Campus dining halls are often supplemented by the foodservice selections available in student unions. The varied offerings include pizza, hamburgers, Oriental, Mexican, and sandwich shops, and these are often found in one central **food court** sharing a main seating area. Some schools also open space for franchise or chain operators, such as Taco Bell and Pizza Hut, to expand food offerings and to better serve students and faculty. Foodservices provided at student unions serve students who may be unable to return to their dining hall for lunch, as well as those who are seeking an alternative to the choices offered at other places on campus. Some schools are adding convenience stores to complement their other food offerings. Students can use their meal cards, or purchase debit cards, to buy food, snacks, soda, paper towels, and other convenience items. The schools are thus able to increase sales while keeping the students on campus to buy these necessities.

Providing food for college and university students is a difficult task. It is nearly impossible to satisfy all students, most of whom have grown up accustomed to one person's way of cooking. Students' diverse backgrounds and life-styles further complicate the matter. Because of all the options available to students, it is difficult to forecast the amount of food to produce. Providing diversity in the menu is another problem for college and university foodservice providers—anyone quickly tires of eating three meals a day produced at the same place, even in his or her favorite restaurant.

Employee Dining Operations

Many corporations provide foodservice to their employees as a perk or as part of a benefit package. In-house restaurants offer two advantages to employees: convenience and cost savings. The cost savings over comparable restaurants outside the company are diminishing, however, as corporations are severely cutting or totally eliminating subsidies to these programs and forcing the food operations to be self-sufficient. Still, the advantage of dining in a central dining facility in or close to the office, rather than fighting crowds or traffic, is appreciated by many employees.

Employers were recently forced to revamp foodservice subsidy programs for two reasons. Tightened financial situations in many companies caused some to view meal subsidies as a luxury that could be eliminated, and changes in the federal tax code designated meal subsidies as a taxable benefit and made them hard to manage. In the face of reduced subsidies, companies now often try to make these food and beverage operations profit centers. Some companies are adding complementary services such as catering, takeout, and even dry cleaning and day care in order to increase revenues and make up for lost subsidies.

School Foodservice

Nutritional deficiencies discovered in American soldiers during physical examinations prior to U.S. entrance into World War II prompted the National School Lunch Act of 1946. The current federal school foodservice program provides hot lunches to about 24 million students daily. The price students pay for meals is prorated according to their families' financial situation. Schools in poorer districts also provide federally subsidized breakfasts for students to increase their concentration level and improve their ability to learn.

Hospital Foodservice

Providing **hospital foodservice** poses challenges not faced by other types of foodservice operations. The diversity of people being fed in a hospital ranges from visitors to staff and professionals, who are on duty around the clock, and patients on various diets with various nutritional requirements. Moreover, a majority of the meals served must be delivered to patients whose rooms are generally quite a distance from where the food is prepared.

Hospital foodservice is also being affected by a change in hospital occupancy patterns. Hospitals are offering more medical procedures on an outpatient basis, which limits the length of time patients stay in the hospital and, in turn, reduces revenue for the foodservice department. Because fewer patients are staying overnight in hospitals, the number of needed meals is reduced, thereby causing underutilization of hospital kitchens that were designed and equipped to provide meal service for a greater number of patients. Many hospitals' foodservice departments are seeking alternative sources of revenue, such as catering and providing food for takeout.

TRENDS IN FOODSERVICE FOR THE 1990s

Trends in American society are reflected by trends in foodservice. Historically, the foodservice industry has adapted to changes in our country. The problems of labor shortages, the state of the economy, the saturation of landfills, and the continuing concern about personal health—all are having a dramatic influence on the foodservice industry of the 1990s.

NATIONAL DEMOGRAPHIC TRENDS

The **demographics** of the United States is changing. Population growth is slowing, which means fewer workers and increased competition for those workers among businesses. The number of young Americans aged 16 to 21 is the lowest in many years, which poses a problem, because this is the age group that provides a large percentage of workers for the foodservice industry. Many operators are turning to the use of automation and convenience foods to ease the problem. Creative operators are also tapping the senior citizen population for workers to fill the labor shortage.

America is also becoming more racially, ethnically, and culturally diverse. This shift in the makeup of the nation's population will affect both what foodservice operations serve and who constitutes their work force. Training programs will be needed for a growing number of non-English-speaking employees. Managers will have to become acquainted with the languages and cultural values of an increasingly multicultural work force.

FOOD TRENDS   *Whats Popularity?*

The NRA's monthly magazine, *Restaurant USA*, highlights the areas of *service, food,* and *casual but high-quality dining experiences* as three aspects of dining that foodservice patrons will be seeking in the 1990s. The NRA states that *service* means focusing more attention on details, providing better service for the more demanding customer. In regard to *food*, the NRA recommends a return to the basics: simple and nutritious meals. Finally, the NRA notes that customers continually place *quality* at the top of the list of priorities for a good foodservice operation, thus the emphasis on *casual but high-quality dining experiences* as the third area highlighted. Most customers tend to prefer a casual, rather than a stuffy or pretentious atmosphere. Foodservice operators who heed these recommendations will continue to succeed.

The increasing level of competition among foodservice operators in the nation will be a challenge. Consumers have grown tired of the poor service of the 1980s, and they now demand better. Foodservice operators will be forced to concentrate much more on satisfying their customers if they expect to keep their businesses and grow.

Menus will continue to change with customers' tastes and lifestyles. Their health concerns also continue to guide menu offerings; people are demanding more healthful choices when dining out. Operators are responding by offering decaffeinated coffees, mineral waters, low-fat items, lighter sauces, and fresh fruits and vegetables. Salads remain popular as side dishes and as main course items for customers looking for a lighter meal. Supermarkets, increasing their offerings, will continue to attract customers. The takeout market will continue to grow, as customers prefer to have "someone else do the cooking" yet still eat at home. Both takeout and convenience food businesses are increasing in order to serve people on the run.

SUMMARY

Today's foodservice industry grew out of a need to provide lodging and, later, meals to travelers. As travel increased, the industry grew. Then, as people began to settle in cities, the need for lodging decreased and restaurants grew in popularity. With the development of cities and towns, competition among foodservice operations forced some out of business and caused others to increase their level of service to please their guests and to encourage repeat business.

The foodservice industry includes more than restaurants. Although restaurants are its largest segment, this vast industry includes all service of food outside the home. The industry is divisible into two major categories: commercial and noncommercial. Within each category are a number of subcategories. All segments of the industry share some common core characteristics, and each has some distinct differences.

KEY TERMS

tavern
American plan
European plan
gross national product
franchise
franchisor
franchisee
restaurant chain
independent ownership
sole proprietorship
entrepreneur
National Restaurant
   Association (NRA)
commercial, for profit
bottom line

full-service restaurant
average check
fast-food restaurant
quick service
commercial cafeterias
steam tables
social caterers
food contractor
lodging places
room service
grocery store foodservice
convenience stores
noncommercial
college/university
   foodservice

captive audience   school foodservice

food court   hospital foodservice

National School Lunch Act   demographics

DISCUSSION AND REVIEW QUESTIONS

1. How has the foodservice industry adapted to the needs of its customers through the years? Cite two examples from the past and two modern examples.

2. What is the origin of the modern diner-type restaurant? Why do you think there is a resurgence of this type of restaurant in several metropolitan areas?

3. Explain the advantages and disadvantages of each of the three forms of ownership in the foodservice industry.

4. What are the advantages and disadvantages of both limited-menu operations and full-service operations, from the perspectives of both the customer and the business owner? Give examples of both types of operations.

5. What are food contractors? Why would a business or a school hire one to assist in its foodservice?

6. Explain why it would be advantageous for a person who is primarily interested in the lodging aspect of the business to learn the principles of foodservice. Discuss the various types of food and beverage outlets in a lodging facility.

7. Why has the growth of foodservice in convenience and grocery stores outpaced its growth in other areas of the foodservice industry? Is this growth expected to continue? Why?

8. What is the difference between a for-profit and a not-for-profit foodservice operation? Discuss two of the types of businesses that operate in the noncommercial segment.

9. How has college foodservice changed and adapted to the times? Cite three examples. Name two things that make college foodservice difficult.

10. The NRA highlights three areas of importance in foodservice for the 1990s. How can a foodservice operation meet these challenges? Give an example of each.

SUGGESTED READINGS

Khan, Mahmood. *Concepts of Foodservice Operations and Management*, 2nd ed. (New York: Van Nostrand Reinhold, 1991).

Lattin, Gerald. *The Lodging and Foodservice Industry*, 2d ed. (East Lansing, MI: Educational Institute of the AH&MA, 1989).

Lundberg, Donald. *The Hotel and Restaurant Business*, 5th ed. (New York: Van Nostrand Reinhold, 1988).

McCool, Audrey C. *Inflight Catering Management* (New York: John Wiley & Sons, 1995).

McCool, Audrey, Fred Smith, and David Tucker. *Dimensions of Noncommercial Foodservice Management* (New York: Van Nostrand Reinhold, 1994).

Powers, Tom. *Introduction to the Hospitality Industry*, 3rd ed. (New York: John Wiley & Sons, 1995).

Shaw, Russel. "Strictly Business." *Restaurants Hospitality* (June 1990): 180–184.

Warner, Mickey. *Noncommercial, Institutional and Contract Foodservice Management* (New York: John Wiley & Sons, 1994).

Weinstein, Jeff. "Looking into the '90s." *Restaurant & Institutions* (27 November 1989): 22–24.

West, Bessie, and Levelle Wood. *Foodservice in Institutions*, 6th ed. (New York: Macmillan, 1988).

CHAPTER TWO

THE ROLE OF THE CUSTOMER

CHAPTER OBJECTIVES

After reading this chapter you should be able to:

- Describe the importance and the implications of service in the foodservice industry.
- Discuss the importance of the customer in a foodservice operation.
- State the problems caused by customers not voicing their dissatisfaction.
- Examine the importance of making an operation customer-driven.
- State four reasons to make an operation customer-driven.
- Distinguish between the characteristics of goods and services.
- Recognize the various components of the customer–service transaction.
- Improve customer service by understanding the importance of the customers' anticipation of the meal, the actual experience, and their feelings when they leave the restaurant.
- Examine when and why people dine out at fast-food and moderately priced restaurants.
- Describe some of the things with which customers may be satisfied when dining out.

In the last few decades the economic base of the United States has shifted from industry and manufacturing to service-oriented businesses. The result has been a change in emphasis, from how well we as a nation produce things to how well we perform tasks. This rather abrupt switch in emphasis within the business world has caused many problems and may be partially to blame for the reduction in the competitiveness of American companies in world markets.

The poor state of "service" in America has become the subject of jokes and the topic of many editorials and magazine articles. Partly to blame may be the fact that service positions in foodservice operations are most likely to be filled by poorly motivated employees at the lowest rung of the salary scale. For service to improve, employees must be made aware of the great importance of customers to their business—and to their jobs. They must also be given the training and tools needed to perform their jobs.

The purpose of this chapter is to provide future foodservice managers with the necessary information to better understand their customers. A satisfied customer is more likely to return. Typically, most of the training managers receive covers the procedural points of running the business and is often inadequate in stressing those skills that add up to good service. Managers and owners often lose sight of the ultimate objective of a service operation: satisfying the customer/guest.

THE IMPLICATIONS OF SERVICE IN THE FOODSERVICE INDUSTRY

> Life is service. The one who progresses is the one who gives his fellow men a little more—a little better service.
>
> —Motto of E. M. Statler, Hotelier

Service plays an important role in the foodservice industry, as illustrated by the fact that *service* is included in the industry name. It is the foundation of the foodservice industry. Customers seek service when they dine out; if they desired a meal without service, they could simply buy food at a grocery store. Customers are willing to pay a premium for the service provided along with the cost of their meals.

Service is the buzzword of the decade. When economic times are tight, customers become more conservative about their spending. Most view dining out as a luxury rather than a necessity. Thus, in difficult economic times most people demand more value for their money, or they try to make their dollar stretch further. To please the increasing number of value-conscious customers, businesses must meet or exceed their patrons' expectations of service.

A National Restaurant Association (NRA) survey found that 90% of restaurant operators believe that their customers are satisfied or very satisfied with their service. This figure differs significantly from the result of a Gallup survey that found that only 60% of customers said they were satisfied with the service they received while dining out. This difference between customer and operator perceptions of the level of service provided illustrates a problem that plagues the foodservice industry: customers do not voice their dissatisfaction about their meal or service. Instead, they just do not return!

A 1986 study by the U.S. Office of Consumer Affairs found that between 37% and 45% of all customers who are dissatisfied with service *do not* complain, and that 30% of customers who are dissatisfied and who do not complain do not return to patronize the business. Customers who do not voice their complaints, thereby giving management an opportunity to remedy the problem, are unfair to the foodservice operation. They make the job of ensuring customer satisfaction more difficult because they condemn the operation without giving it a chance to correct the difficulty. The management and service staff of a foodservice operation are therefore forced to become more observant of their customers' nonverbal signals in order to detect problems and correct them to ensure that the customers will return.

The foodservice industry employs many people in a variety of positions, both **front-of-the-house** (the parts of the operation that customers see and interact with) and **back-of-the-house** (the parts that customers do not generally see). One problem is that most employees—possibly because they have not been trained to do so—do not view their jobs and roles as "customer pleasers." Instead, many people in customer-service positions view customers as pests or "not my job," rather than as the providers of the capital that keeps the restaurant in business and provides the money for their paychecks. An example of this attitude is the general reaction of the staff of a restaurant when a customer arrives 10 minutes before closing. The customer receives rushed service from servers who would rather be cleaning up so they can go home, and mediocre food because the cooks are concerned with going home and have put away all the food and shut off the ovens and stoves. The result is a dissatisfied customer who will probably never return to that restaurant.

THE CUSTOMER-DRIVEN OPERATION

The definition of **customer**, according to *Webster's New Collegiate Dictionary*, is "one that is a patron (as of a restaurant) or that uses the services (as of a store)." In other words, a patron or customer is someone who exchanges his or her money for goods and services. Customers' patronage of a foodservice operation provides the money that is the lifeblood of the business. They provide the money that pays the bills, the payroll, and generally keep the doors of the business open.

The dictionary definition of *customer*, however, does not adequately stress the importance of customers to a foodservice operation. To attract and then retain customers, a foodservice operation must view them in a broader perspective. They must be treated as the most important component of the operation. The entire operation must be designed around the customers and their needs. A foodservice operation must, in fact, be **customer-driven**.

WHY MAKE AN OPERATION CUSTOMER-DRIVEN?

Foodservice operations *must* make their operations customer-driven for the following reasons:

1. *To distinguish the operation from its competition.* Customers remember good, attentive service. In similar types of foodservice operations (for example, Burger King and McDonald's) prices are generally similar and the choice and quality of products they offer are also generally similar. With all other factors being relatively equal, then, the main reason people usually choose one place to dine over another is that one does a better job of meeting their needs, whether it be through the speed or effectiveness of the server, the friendliness of the host or hostess, or the efficient way a complaint was handled on a prior visit. When customers leave an operation, the level of service they received is generally what they will remember most.

2. *To build the operation's market share, or percentage of the market, as compared with other operations in the area.* **Market share** is first gained by offering a product different from any offered by a competitor. This accounts for the fact that most new restaurants are busy when they first open—they are offering something different that customers seek. Once the competition

catches up and matches the unique offering of the new place, the differences between the operations are equalized. Therefore, excellent service and an effort to retain loyal customers will allow the customer-driven operation to increase its share of the market.

3. *To build customer loyalty*. Customer loyalty and profitability go hand in hand. Repeat guests are the best advertisement—they bring in new customers for the operation through positive "word of mouth." Negative comments, obviously, have a detrimental effect. Prospective customers are more likely to believe what their friends say about an operation than what they read in an advertisement. Repeat customers know the operation's menu and services; therefore, they are generally easier to serve.

4. *To identify potential mistakes*. Preventing mistakes will help in serving the customer better and will save time and money for the operation. All employees should become "quality inspectors." The little extra time spent checking can help in avoiding mistakes before they happen.

WHY MAKE AN OPERATION CUSTOMER-DRIVEN?

1. To distinguish your operation from your competition
2. To build market share
3. To build customer loyalty
4. To identify mistakes before they happen

HOW TO MAKE AN OPERATION CUSTOMER-DRIVEN

Make Customer Service a Commitment

Where does an operation begin in making its service customer-driven? Focusing an entire organization on taking care of its customers means more than the mere declaration of a new strategy or policy. It requires a commitment of all the company's resources and personnel. It involves evaluating company priorities and management and staff attitudes, behaviors, and policies.

Improving and maintaining high service standards should not be viewed simply as a goal to be met, but rather as an ongoing and unending process. The ultimate measure of an operation's success in satisfying its guests is also the most effective form of advertising—word of mouth. It is the customers who will judge whether the program is successful, as they are the people who must ultimately be satisfied.

Each operation must make a commitment to customer service. Quality service should be viewed as more than just a passing fad; in an ever more competitive business environment, it is a matter of survival. All employees of an operation, from top management to the hourly staff, must be aware of the new focus of the company. Commitment must start at the top and filter down. The whole process will collapse if top management, rather than becoming totally involved, only mildly supports the program.

Many managers have the misconception that efficient service of high quality is too costly to implement; therefore, they do not set such a standard. This could not be further from the truth: quality service does not cost, it pays. Benefits are gained both in tangible ways, with the improvement of sales and profit, and in intangible ways, through boosted customer goodwill and the increased morale of the staff.

Pay Attention to Customers

A recent Massachusetts Institute of Technology study found that 80% of all technological innovations came from customer suggestions. Managers and operators need to ask the people, whose satisfaction is the goal, what they can do to provide better service. Rather than guessing about what customers want, it is best to go right to the source.

Because customers generally do not voice their satisfaction or dissatisfaction, employees must be aware of actions that indicate customers' feelings. For example, food that was not eaten or that was barely touched could signal a problem, warranting a question from the server as to why the food item was not eaten.

Encourage Customer Suggestions

Front-of-the-house staff, especially servers, spend more time than other staff with the operation's customers and can be an excellent source of information about customer comments. Back-of-the-house staff, especially dishwashers, can report any unusually large amounts of leftover menu items that are returned to the dish area. Two possible reasons for leftover food are poor food quality and portions that are too large.

Any information obtained from the customers, through either direct observation or conversation, must be communicated to management and the other staff members who have an interest in such knowledge. A clear path of information must flow freely between all of the people involved, including management, preparation staff (cooks, chef), utility staff (dishwashers, pot washers), service staff (waitresses, waiters, bus people, host or hostess), and customers. For example, if there are many customer complaints than the soup is too salty, these should be relayed to the preparation staff and management so that the problem can be resolved. If the server does not pass on the information, the problem will not be taken care of, and the customers who subsequently order that soup will also be dissatisfied. The timely passing of information by all staff members is crucial to the detection and correction of problems so that customers can receive the highest possible quality of service and food.

Recognize and Reward Good Service

In order to keep employees committed to providing good service, they must be recognized and rewarded. Customer service should be included in the evaluation criteria for all staff. Customer comments and direct observations provide valuable information in judging the customer service that employees are providing. Employees like to be recognized among their fellow workers with a certificate or gift, but a formal ceremony is not necessary. Some restaurants trade certificates so that they can offer employee dinners at other dining places at minimal cost.

HOW TO MAKE AN OPERATION CUSTOMER-DRIVEN

1. Make customer service a commitment.
2. Listen to and pay attention to the customers.
3. Recognize and reward good service.

CHARACTERISTICS OF GOODS AND SERVICES

There are actually two components combined in a menu item that a customer orders and purchases when he or she visits a foodservice operation: **goods** and **services** (See Figure 2.1). To

Restaurant

FIGURE 2.1 Goods and services.

better understand the concept and importance of making an operation customer-driven and how an operation can better serve its guests, it is necessary to examine these components.

Goods (the actual food or drink) and services (the preparing, cooking, and delivery of the products that are included with the food or drink) are generally inseparable. Diners at a foodservice operation are looking to purchase more than just food; they come to buy a dining experience. Most customers are willing to pay at least two to three times the cost of the food at a grocery store for the service provided with it. Although the quality of the goods is important, it is the service that usually leaves a more lasting impression with the customer and can make or break the dining experience.

Service is intangible; it is something that cannot be physically touched or grasped. In contrast, goods are things that can be handled or consumed. To illustrate the difference, consider the hamburger: the goods are the hamburger, bun, lettuce, and tomato—all of the things you can pick up and eat. The service, however, is the preparation of the bun, the lettuce, and the tomato, the cooking of the hamburger, and the delivery to the guest. The two are inseparable: the hamburger (goods) cannot be presented without the second component (services).

In a foodservice operation, service requires goods to make the transaction complete. Restaurant servers cannot perform their jobs without food or drink to serve to their guests. One thing that distinguishes one operation from others is how the goods and services provided are combined.

GOODS

The goods that foodservice operations sell include food and beverages—those things that customers eat and drink. The quality of the food served is important to customers. Foodservice operations have options as to the quality of the items they serve—either name-brand items with customer recognition, such as Heinz Catsup and Pepsi-Cola, or items with brand names that customers do not recognize. Most foods sold and served in commercial operations do not have names that customers are aware of or recognize from grocery stores or advertising.

SERVICES

Although the quality of food served is important, the level and quality of service provided will have a more lasting effect on most customers. All other factors being equal, a mediocre meal of lesser quality, but presented with superb, attentive service, will result in a satisfied guest. In contrast, an excellently prepared meal made of high-quality ingredients, but with poor service, will result in a dissatisfied guest.

Service is also important because of its crucial role in determining the customer's perception of the value of the meal. The higher the level of service provided with the food, the more the customer is generally willing to pay. Consider, for example, the significant difference between the price of a hamburger at a self-service fast-food operation and the price of a similar item at a table-service restaurant.

As a further example, compare the purchase of a pizza at a food court in a mall with buying pizza from an upscale pizza restaurant. Most customers would not expect much, if any, service from the food court in the mall. There is generally little choice of toppings or accompaniments, and one might have to settle for whatever slices are in the warmer. The two pizzas are basically the same: crust covered with sauce and toppings. However, customers at the upscale restaurant have greater expectations and are willing to pay a premium for the increased service provided. They will expect a pleasant atmosphere, comfortable seating, accommodating service, menu suggestions, and a variety of topping and accompaniment choices.

A 1991 NRA survey indicated the importance customers place on service. Both *timely service* and *friendly staff* were rated in the top 25% of characteristics determining good value in a moderately priced, table-service restaurant.

COMPONENTS OF THE CUSTOMER–SERVICE TRANSACTION

To be successful in satisfying guests, managers must focus on the interactions between their guests and the various components of their operations. Once managers understand each step in the transaction, they can better serve their guests. The exchange that occurs between a guest or customer, and the service operation comprises three steps (see Figure 2-2):

1. The **anticipation**, or expectation prior to visiting the restaurant,

2. The **actual experience**, or what happens while dining, and

3. The **impression**, positive or negative, with which the customer leaves.

ANTICIPATION

Before actually dining at a restaurant, customers form various expectations about the meal or dining experience they are about to purchase. The expectations will later be compared with the actual service provided. For customers to be satisfied, the service they receive must meet or exceed their expectations. The problem is that customers sometimes have unrealistic expectations about a restaurant, based on advertisements or on what they have heard from other people, which makes them hard to satisfy.

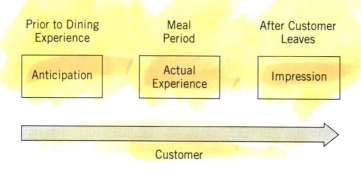

FIGURE 2.2 The customer service transaction.

Customers anticipate the value they will receive for the money they spend; people are generally concerned with what they get for their money, or the price/value relationship. The higher the price paid, the more value is expected. This is of particular concern in tight economic times when customers are trying to make their money stretch as far as possible.

Customers can also be concerned with the utility of the meal provided. They may try to anticipate whether the restaurant will provide enough food to satisfy them and their guests. The question is of particular concern to people planning parties or conferences, where the quality and/or quantity of food will influence the participants' impression of the affair (such as a wedding). Here, the price/value relationship is evidenced: some customers relate the amount of food provided with the cost and value of the meal.

If a customer is patronizing an operation for the first time, he or she may view the upcoming dining experience as a risk. This feeling is amplified if the customer is entertaining friends or business acquaintances and the success of the function (meeting, convention, or wedding) is partly dependent on the quality of food and service provided.

THE ACTUAL EXPERIENCE

The actual experience segment of the transaction comprises both tangible things, such as the quality of the food, and intangible things, such as the atmosphere of the dining establishment. The food is only one of the many factors that influence a customer's dining experience.

The actual experience begins as the customer approaches the establishment. The attractiveness of the setting, as well as the condition and upkeep of the building exterior, is important to the customer's experience of the meal. A sign with burned-out lights and flaking paint or a run-down building may cause the customer to think less of the entire operation.

Once the customer has entered the building, the atmosphere of the operation is evaluated. The background music, the level of lighting, and the type and condition of the interior furnishings are judged and compared with his or her expectations. The service provided by the host or hostess also influences the customer. Customers expect to be greeted promptly, or at least acknowledged, upon their arrival. Those with reservations expect to be seated promptly at their reserved time, not directed to the bar for a cocktail.

The initial contact with the personnel of the restaurant makes a very lasting impression on most guests.

Customers use all five senses to survey and evaluate a foodservice operation. Sight provides the most information. People may make inferences based on what the operation does and does not allow them to see, as illustrated by the saying, "You can judge a restaurant by the cleanliness of its rest rooms." Customers assess the cleanliness of a rest room, an area of the "back of the house" that management lets them see, and draw conclusions about the cleanliness in areas of operation they generally do not get to see, such as the kitchen.

The scents and odors that fill a dining establishment influence customers as well. The savory aroma of food cooking is a welcome delight for diners and does wonders to stimulate the appetite. Some operations place their bakeshops in or near the dining room so that customers can be greeted by the enticing smells of freshly baking items. On the other hand, excessive chemical odors or the smell of stale beer or smoke can quickly create a negative impression.

The temperature of the dining room also affects the customer's enjoyment. A room that is either too cold or too warm can cause the customer discomfort and can detract from the other components of the operation. Because sounds can either complement the experience or diminish it, background music must be chosen carefully. Different people have different tastes in music, and one must be careful not to offend one group while trying to satisfy another with any type of music. The volume of the music is also important and must be carefully considered. Likewise, the noise of an ice machine, a dishwashing machine, or the group in the next booth may interfere with guests who were looking forward to a quiet evening out.

The taste of the food is important and must be monitored closely to please the guests. People have different tolerances in regard to tastes, such as spices or salt. Preparation personnel must remember that a guest can add salt more easily than one can remove it from a food item.

Once the guest is seated at a table, he or she begins to evaluate the actual delivery of the service. The server is the most important part of this segment of the experience. Customers expect prompt and courteous service. The server must be knowledgeable about the menu offerings and able to make suggestions if so requested. Food must be delivered to the table while it is still hot, complete with the necessary condiments and any special eating utensils. It is very annoying to customers for a server to deliver an item such as french fries to the table and then ask if they want catsup. The french fries get cold while they wait for the catsup that should have been brought to the table prior to delivery of the fries. To increase the level of service, the server should return to the table shortly after the food is delivered to remedy any problems and to make corrections quickly. Customers appreciate this attention.

IMPRESSIONS

Customers' feelings and impressions upon leaving the operation will have a lasting influence on whether or not they will return. A "Please Come Back" sign on the door is not enough to bring back customers who were not treated the way they thought they should have been treated. A decision to revisit the operation, or to recommend the place to friends or family, is made once the customer evaluates all aspects of the dining experience. Most customers compare their expectations with their actual experiences. The management and staff of the operation must work hard to make sure that a customer's problems, if any, are resolved and that he or she leaves satisfied. The importance of the lingering impression with which the customer leaves is a priority of a customer-driven operation.

LEVEL OF MANAGEMENT IN THE CUSTOMER–SERVICE TRANSACTION

In each of the three steps in the customer–service transaction, there is a certain level of management control. Management must take care, in each of the aspects it controls, to ensure that customers are treated as well as possible while they are dining. Most managers concentrate on the "actual experience" segment of the service transaction. Although this is important, the other two segments have a critical impact on the customer's satisfaction and the success of the operation. Care must be taken that advertising portrays the operation in accurate terms so that the customer does not have unrealistically high expectations that are impossible to fulfill. Both management and service staff must be well versed in nonverbal cues in order to recognize and remedy customer dissatisfaction, to ensure that guests leave with a good impression. The steps in this transaction are shown in Table 2.1.

TABLE 2.1 Customer–Service Transaction

The Customer	Step in Transaction	Management Level of Control
Prior to visiting the operation	ANTICIPATION	High in some aspects Low in other aspects
Expectations: customer considers Price/value relationship? Utility Will it please his or her guests?		
During the Visit Customer rates Atmosphere Condition of the operation Taste/quality of food Delivery of service	ACTUAL EXPERIENCE	High
Leaving the Business Customer Compares expectations with actual experience Forms a decision on whether or not to return	FINAL IMPRESSION	Some control

Source: Adapted from: M. Barringron and M. Olsen, "A Model for Service Transactions: The Concept of Service in the Hospitality Industry," *International Journal of Hospitality Management,* Vol. 6, No. 3, 1987.

WHEN AND WHY PEOPLE DINE OUT

People dine out for a variety of reasons, for instance, to celebrate a special occasion, a holiday, or a birthday. Some people, away from home while at work, school, or traveling, may find dining out to be more convenient. Others may just enjoy having someone else prepare their food and clean up the dishes and the kitchen for them.

Depending on the customer's reason and motivation for dining out, different expectations and needs must be satisfied. Information obtained on when and why people dine out can also be used by managers to better plan marketing and advertising materials. When customers have expressed certain reasons for dining out, the goal is then to incorporate ways to fulfill these needs in the management, policies, and promotions of the operation. The data provided in Tables 2.2 through 2.5 was accumulated from NRA surveys and is included here to give examples of the information available to managers about the customers who patronize their businesses.

WHEN PEOPLE DINE OUT

According to National Restaurant Association statistics, the most common day of the year to dine out is a birthday. Approximately 50% of consumers celebrate their birthdays at a restaurant. Mother's Day and Father's Day rank as the second and third busiest holidays celebrated in restaurants. The most popular day of the week for consumers to dine out is Friday, followed by Saturday and Thursday (see Table 2.2).

TYPICAL FOODSERVICE CONSUMERS

On an average day, 37% of the American population dines in foodservice establishments, and 21% purchase food for takeout or "to go." Men dine out slightly more than women, 57% as compared with 47%, on a typical day. Men are also more likely to eat out for lunch than women. It is interesting to note that dining at foodservice operations diminishes with the age of the consumer; typically, 68% of 18- to 24-year-olds dine out daily, whereas only 34% of adults over the age of 65 eat out.

According to a Good Housekeeping/Roper survey done in 1991, increasing numbers of mothers are taking their children out to dinner with them. Seventy-two percent of mothers reported that they brought their children with them when dining out and that the children were influential in the decision to dine at a table-service restaurant. Families with children are more likely to patronize quick-service establishments and to take out their food to eat elsewhere. This may be due to the fact that quick-service operations

TABLE 2.2 The Likelihood of Eating at a Restaurant on Various Holidays

	Birthday	Mother's Day	Father's Day	Valentine's Day	Easter	New Year's Eve	St. Patrick's Day	Thanks- giving Day	Christmas
Men	47%	39%	24%	22%	16%	13%	10%	10%	6%
Women	42%	39%	25%	24%	18%	14%	12%	10%	6%
Overall	52%	40%	24%	21%	14%	12%	8%	10%	5%

Source: Adapted from *Holiday Dining*, National Restaurant Association, 1990.

serve food that is more popular with children—pizzas, hamburgers, and Mexican foods—and often serve meals specially packaged with activities and toys that appeal to children.

WHY PEOPLE DINE OUT

People generally dine in fast-food restaurants (restaurants without table service where customers order food at a counter) or in moderately priced table-service restaurants (family-type establishments where table service is provided and the average check is less than $8 a person). The convenience of fast service and an accessible location are prime selling factors for both types of operations. This is a further indication of the increased pace and rushed life-styles of today's Americans. The increased concern for value in tight economic times also fuels the popularity of these types of operations.

Moderately priced restaurants (see Table 2.3) are most popular with middle- to upper-income males, Southerners, and consumers in the middle- to upper-income brackets. These restaurants' increased level of service, as compared with fast-food restaurants, is particularly welcomed by more mature customers who like to have their food served to them rather than having to serve themselves or go to the counter to pick up their food.

Sixty-five percent of customers who dine at fast-food restaurants are very much concerned with time and convenience—again, reflecting today's hectic life-styles. Price is important to only about 8% of adults who visit places of this type (see Table 2.4).

TABLE 2.3 Reasons for Visits to Moderately Priced, Table-Service Restaurants

Reason for Visit	Percentage of Adults Reporting
Night out/weekend out	15
Less expensive	14
Convenient	13
Time factor	12
Family outing	10
Special occasion	9
With friends/guests	8

Source: *Tableservice Trends,* National Restaurant Association, 1992.

TABLE 2.4 Reasons for Visits to
Fast-Food Restaurants

Reason for Visit	Percentage of Adults Reporting
Time factor	48
Convenient	17
Shopping/traveling	12
Take children out	8
Less expensive	8
Do not feel like cooking	8

Source: *Trends in Food*, National Restaurant Association, 1991.

CUSTOMER SATISFACTION

Cleanliness is rated as the highest concern for customers in both fast-food and moderately priced restaurants. The differences in food quality, variety, and atmosphere between the two types of restaurants, shown in Table 2.5, indicate areas that customers view as important and for which they are willing to pay a premium when dining out. The relatively low percentages given for food quality and perceived value in fast-food restaurants indicate customer dissatisfaction in these two areas. **Perceived value**, or the comparison between what the customers paid for their meals and what they thought the meals were worth, is a topic that fast-food operations are addressing by offering value meals, combination meals, and unlimited refills for soft drinks.

Table 2.5 provides crucial information for foodservice managers and is critical to designing and maintaining a customer-driven operation. Remember, before the customer can be satisfied, the foodservice manager needs to know what the customer wants and needs.

SUMMARY

The customer must be viewed as the most important component of any foodservice business. The customer provides the capital and funds for the business to operate. Customers can sense when an operation is not concerned with them or their business. Those who feel that they are unimportant to the operation will take their business elsewhere.

Managers need to understand the implications of service in the foodservice industry, the importance of making an operation customer-driven, the characteristics of goods and services, and the components of the customer–service transaction. By understanding these concepts, as well as the reasons and motivations of customers for dining out, managers can set priorities and goals and more efficiently organize and run their operations.

Improving service does not mean changing recipes or ingredients, although both are important if customers are dissatisfied with them. Improving service is accomplished by realigning the goals and objectives of the operation toward meeting and exceeding guests' expectations. All areas and personnel of an operation, from executives and top managers to staff and support personnel, must be dedicated to the cause. Staff training and recognition must be focused on improving and maintaining a high level of customer service.

KEY TERMS   *need to know*

service	intangible
front-of-the-house	goods
back-of-the-house	services
customer	anticipation
customer-driven	actual experience
market share	impression
tangible	perceived value

DISCUSSION AND REVIEW QUESTIONS

1. What is the difference in perceptions of the level of service provided to the dining public? What factors contribute to this difference in perception? What problems does this difference cause?

2. Why must all employees of a foodservice operation understand that one of their roles is to be customer pleasers? What problems may arise if they do not?

3. Why is the dictionary definition of the word *customer* inadequate as it relates to foodservice operation? What is the meaning of the word in foodservice terms?

4. Explain the four reasons for making an operation customer-driven.

5. Explain why foodservice personnel at all levels must be committed in order to achieve success in making an operation customer-driven.

6. Explain the relationship between goods and services in a foodservice operation; also explain the concepts *tangible* and *intangible*.

7. What are the three components of the customer–service transaction? Explain which parts of the transaction managers control, and which parts they do not control. What problems can occur with the parts of the transaction that management cannot control?

8. Explain why a "Please Come Back" sign on the door is not sufficient to bring customers back.

9. Why is it important for foodservice managers to understand the reasons that customers dine out?

10. Suppose you are the manager of a moderately priced, table-service restaurant. Based on the reasons given in this chapter on why people dine out, outline some suggestions to improve business. Then make suggestions for improving business as if you were the manager of a fast-food establishment.

SUGGESTED READINGS

Barringron, M., and M. Olsen, "A Model for Service Transactions: The Concept of Service in the Hospitality Industry." *International Journal of Hospitality* (1987), Vol. 6, No. 3.

Davidoff, D. *Contact: Customer Service in the Hospitality and Tourism Industry* (Englewood Cliffs, NJ: Prentice-Hall, 1994).

Marvin, B. *Restaurant Basics* (New York: John Wiley & Sons, 1992).

Sullivan, J., and P. Roberts, *Service That Sells!* (Denver, CO: Pencom Press, 1991).

CHAPTER THREE

FOODSERVICE ORGANIZATION AND SYSTEMS

This chapter focuses on three aspects of the foodservice industry that are of primary importance to a potential manager: the major classifications of foodservice systems, foodservice organizational theory, and the levels of personnel involved within the organizational structure.

The four major classifications of foodservice systems are based on a number of factors, such as the types of food purchased and the amount of time that elapses between the time food is prepared and when it is served. To best understand the foodservice industry, a potential manager should be aware of the advantages and disadvantages of each of the four systems.

An organization is a group of people united to achieve a common goal. One of the roles of a manager is to assign tasks within the organization; therefore, an effective manager must understand the basics of organizational structure. The labor-intensive nature of the foodservice industry also makes it essential for managers to understand how to increase and maintain worker productivity.

The foodservice industry comprises three basic types of workers: managers, production personnel, and service personnel. Within the position of manager there are three levels: top, middle, and supervisory. The roles and responsibilities of each type of worker and manager are outlined in this chapter.

TYPES OF FOODSERVICE SYSTEMS

Changes and innovations in foodservice systems in recent years have been driven by the need for greater labor efficiency owing to steadily increasing labor costs and the foodservice industries' need to adapt to the changing dynamics of their clientele. The *conventional system* (also known as the traditional system) is most common to the majority of restaurant-type operations. The *commissary system* developed from the growth of chain restaurant operations and the need to consolidate some of the food preparation procedures. Ready-food systems were developed in an attempt to smooth out the abrupt "peaks and valleys" of foodservice production. The *convenience food system*, whereby the food needs only to be heated, serves busy people who want to grab a quick meal at such places as convenience stores, grocery stores, and gas stations, rather than in a conventional restaurant. For an overview of the four foodservice systems, see Table 3.1.

CONVENTIONAL SYSTEM

The **conventional** or **traditional foodservice system** operates similarly to the way people cook at home. The system is characterized by preparation of most food when it is needed, or "to order," using a minimum of preprocessed foods, or from "scratch." In the past this was the most common method of preparing food for commercial operations.

The food items in a conventional operation are generally purchased in their most basic form with a minimal amount of processing. Menu items are prepared as needed or ordered and are then held, either hot or cold, for the shortest possible amount of time

TABLE 3.1 Foodservice Systems

System	Characteristics	Type of Operations
Conventional/Traditional	Similar to cooking at home.	Independent restaurants
Commissary	Food preparation is started in one place, then preparation is finished and food served by satellite kicthens.	Chain restaurants
Ready-Food	Food is prepared in batches and reheated as needed.	Hospitals, restaurants
Convenience Food	Food is prepared and packaged in one place and sold, ready to eat, at another location.	Convenience and grocery stores

before being served. Food is prepared in the smallest quantities possible. An advantage to the conventional system is that the customer is treated to a meal made from fresh food without preservatives or unnecessary additives. Potential disadvantages are that such a meal may be prohibitively expensive, because of the great amount of labor required to prepare it, and it may take longer to prepare than most people find convenient.

An additional advantage of the conventional system is that it allows for greater creativity and flexibility in menu items. Because all items served are prepared on the premises, the kitchen staff has an opportunity to create as it wishes and to customize items. The opportunity to customize dishes allows a restaurant to develop a uniqueness that may attract more business. A major disadvantage to making all menu items from scratch, in addition to the amount of labor required, is the potential for a lack of consistency in food products from day to day and from cook to cook.

Because most of the menu items served are **prepared "in-house,"** the conventional system is very labor-intensive, meaning that it requires a large number of labor hours to produce. Further complicating the problem is the cost of skilled kitchen personnel, which has skyrocketed owing to dwindling availability. In addition, an operation using the conventional system requires a highly skilled staff and a large kitchen filled with a great variety of food-production and food-storage equipment.

The conventional/traditional system has been modified to fit the times. Foodservice operations still prepare a portion of the food items they serve **"to order,"** but they do not purchase all food items in the unprocessed state. Operations today generally purchase their food items in various stages of processing. The increased use of processed goods allows an operation to maintain a variety of items on the menu while using fewer preparation personnel and less skilled labor.

COMMISSARY SYSTEM

A **commissary** is a central food production facility that prepares food in large batches to be sent to satellite, or remote, kitchens to be heated or finished and served. A commissary generally prepares most items from unprocessed ingredients and a minimum amount of convenience goods. A major advantage of using the commissary system is that a foodservice company can be fairly well assured that the product will be consistent for all of its customers, regardless of location.

Envision a company that operates three to five soup and salad bar restaurants within a 50-mile radius. If the operation used the conventional system, each restaurant would need its own trained kitchen staff and a fully equipped kitchen. Each day, in each restaurant, the kitchen staff would have to prepare the soups and dressings for the day's business. From day to day and from operation to operation, consistency could not be guaranteed. Each operation would be buying the ingredients and supplies it needs in small quantities.

To differentiate between these two systems, consider the operation of a soup and salad bar restaurant chain that uses the commissary system rather than a conventional system. Instead of each operation needing a trained food-preparation staff and a fully equipped kitchen to prepare the soups and dressings for the day, it requires a less skilled person and a reduced amount of equipment to heat and serve the items delivered from the commissary. The food items are shipped to each satellite operation daily. The commissary can also purchase its food items in quantities larger than would be feasible for individual operations, thus lowering the cost per unit of the ingredients, which in turn lowers the cost of the menu items. The soups, dressings, and other items that are prepared in one batch are consistent from one operation to the next. The commissary can use highly automated preparation equipment, producing large batches of food with a minimum of labor.

The advantages of a commissary system may seem overwhelming, but a few disadvantages must also be considered. The numbers of needed kitchen preparation personnel may be decreased, but a commissary system has to maintain trucks and hire drivers. What happens if a satellite operation runs out of an item that it does not prepare on the premises? It will not be able to make more. There is also the possibility of a truck failure or accident, leaving the operation without food to serve. A strong communication link must be established between the central commissary and its satellites to monitor the amounts of food to be shipped, reduce waste, and avoid running out of food items.

READY-FOOD SYSTEMS

A major problem in any foodservice is the erratic pattern of meal service. Foodservice operations go through a series of "peaks and valleys," or high and low periods of business and activity, throughout the day, corresponding to meal periods (see Figure 3.1). Ready-food or ready-prepared systems attempt to flatten out this fluctuation. The food preparation personnel prepare menu items in large batches, then chill or freeze it in small batches for serving later. Food production staff remain busy and, therefore, more productive.

Two examples of the **ready-food system** are *sous-vide* (pronounced "sue-veed") and cook-chill. These are used primarily in institutional foodservice operations but have also been used in restaurant foodservice. The advantage is that preparation personnel increase their productivity by producing a large volume of food ahead of time. Food items can be heated and served at any time, without the involvement of trained preparation personnel.

Sous-vide is currently a popular form of ready foods. *Sous-vide*, French for "in vacuum," was developed in the 1970s by a French chef

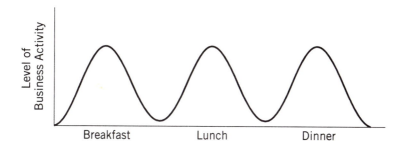

FIGURE 3.1 Peaks and valleys in restaurant day.

and food scientist. In America this is a USDA-regulated process that must be performed under the close scrutiny of federal regulators. Food items, usually individually portioned, are placed in special plastic pouches from which the oxygen has been removed and are then fully cooked at low temperatures. Once a product is cooked, the pouch is rapidly chilled or frozen and held for reheating.

In *cook-freeze* or *cook-chill* systems, foodservice operations prepare food and chill or freeze it to be heated later and served. Foods are generally produced in large amounts and then portioned into smaller units. Foods prepared in cook-chill systems are cooled in a blast chiller that reduces their temperature quickly to 33°–37° F for service in one or two days. Foods prepared for the cook-freeze system are frozen in a blast freezer to below freezing and can be stored from two weeks to about three months. Both processes require special equipment for the operator to reduce the temperature of the food quickly in order to reduce the growth of bacteria.

There are several differences between *sous-vide* and cook-freeze or cook-chill systems. *Sous-vide* is processed to reduce shrinkage. Because *sous-vide* products are always fully cooked and usually packaged individually, the original qualities of the food are preserved. With *sous-vide*, companies have the option of producing their own products or having USDA-approved companies produce items for them according to their specifications. Proponents of the process claim that it has three benefits: convenience of preparation, high-quality finished products, and reduction in the amount of labor needed. *Sous-vide* products can be either flash frozen or held chilled until they are heated for service.

Food processed in *cook-freeze* or *cook-chill* systems do not have to be sealed in plastic packets or have the oxygen removed, and the food does not fall under the scrutiny of the USDA. However, the addition of processing equipment that is very expensive and cost prohibitive for most small operations is required.

CONVENIENCE FOOD SYSTEMS

In a **convenience food system**, food is prepared at a central location and then shipped elsewhere to be sold. An operation that offers food for sale that needs only to be heated is part of a convenience food system. Examples include the foodservice at convenience marts, at gas stations, and at some grocery stores. Food choices are generally restricted to popular food items that can be reheated, such as pizza, sandwiches, and burritos.

Operations using convenience food systems are among the fastest growing segments of the foodservice industry. People appreciate the convenience of having prepared food available for them to eat when they are in a hurry. Grocery stores that sell convenience foods offer alternatives and competition to area restaurants by providing food for people who do not want to cook or, for whatever reason, do not want to dine at a restaurant. Customers are willing to pay extra for the added convenience and speed of this system.

ORGANIZATIONAL STRUCTURE OF FOODSERVICE OPERATIONS

As mentioned earlier, an **organization** is a group of people united to achieve common objectives or goals. This section of the chapter presents some basic principles of organizational theory. Principles that can help organizations to be more effective in realizing their goals are also discussed.

The foodservice industry is considered to be very **labor-intensive**. This is due in large part to the lack of automation and to customers' preference for personal contact while dining out. An effective foodservice manager must understand how to organize, motivate, and work with staff members. The success of a manager in a foodservice operation depends heavily on how well he or she manages the staff. The better a manager understands how people work and what motivates them, the better and smoother the manager's performance. The person in a manager's position who is an expert in preparing or serving food, but unskilled in managing people, will not be as effective as one who is. The following paragraphs also examine various organizational structures of foodservice operations, along with the roles and responsibilities of the various positions.

ORGANIZATIONAL THEORY

Organizational, or management, theory has evolved quite a bit during the last century. Prior to the 1900s, most work was done in small shops or factories by small numbers of workers. The population of the nation had not grown much, and most people lived and worked on farms. At the turn of the century immigration brought many people to the "land of opportunity," and the Industrial Revolution began. More people meant that more consumer goods were required to be manufactured in greater numbers. As small shops grew to large factories, the need for the study of organizations increased. At first workers were seen as tools of production, similar to machines without feelings or needs. Yet as organizational theory evolved, the Human Relations School of thinking emerged, which postulated that if workers were treated as "people" they would be more productive.

Scientific Management

Before the early 1900s management was not a field of study. Frederick Taylor, a steel company foreman, was the first to formulate theories on management and industrial efficiency. Taylor felt that the industrial inefficiency of the time was due to the lack of systematic management and thus developed four principles of Scientific Management, which established the importance of the following steps:

1. Selecting workers according to personal character, habits, and ambitions,
2. Motivating workers to perform,
3. Training workers systematically, and
4. Analyzing the particular task to be performed and altering the work process to maximize performance.

Analysis of a task and changing the work process to maximize performance reflects the principle that best characterizes Scientific Management. Prior to this period, production was not structured to maximize the performance of workers. The use of this principle revolutionized the industrial work of the time, drastically increasing productivity. The Scientific Management method concentrated on maximizing efficiency, applying detailed and uniform work procedures, and standardizing detailed accounting procedures.

Classical School

Henri Fayol developed methods to increase the productivity of workers in France at about the same time that Taylor was working in the United States. The Classical View, as it was called, complemented the Scientific Management method. Classical organizational theorists developed a fixed set of principles that established the basis for management. Elements of an organization were grouped according to function or other similar criteria. The structure of an organization was based on formalized organizational charts so that those involved could determine the relationships between the various groups and divisions.

Fayol described what he thought should be the five duties of a manager:

1. Planning
2. Organizing
3. Commanding employees
4. Coordinating activities
5. Controlling performance

Fayol also formulated 14 principles of management. These are usually reduced to five features crucial to effective organizations:

1. *Specialization*—Work is divided according to logical components, such as product, expertise, place of work, etc.
2. *Unity of command*—Everyone in an organization should have only one supervisor to report to. This is an attempt to limit confusion in growing organizations.
3. *Scalar chain of authority*—The line of command is clearly defined with a formal organizational structure or chart. Responsibility and authority should flow, in the most direct and unbroken path possible, from the top to the bottom levels of the organization.
4. *Coordination of activities*—Lines of communication are established to link all key areas of the organization.
5. *Controlling performance*—Production of workers must be monitored to ensure that workers are working up to performance standards.

Fayol believed that a trained management team was essential to improving the operation of an organization. He was a strong advocate for management-training programs to develop the skill levels of individuals. He focused his attention on the manager, rather than on the worker as Taylor did.

Both the Scientific Management method and the Classical School stress the structure and mechanical nature of organizations; a shortcoming common to both is that the formalized nature of organizations does not consider the "human relations factor." Both systems treated the worker as just another tool of production, similar to a machine, and disregarded the human factor and how it affected the operation of the business. In 1924, a group of researchers began examining how the human factor figures into the worker productivity equation.

Human Relations School

The first studies to consider the human factor in relation to productivity were performed in a manufacturing plant. A study was developed to measure the effect of lighting level in the plant on workers' productivity. It discovered that productivity increased, or at least remained the same, regardless of whether the level of lighting was increased or decreased.

The experiment provided the first connection between workers' feelings and their productivity. When employees' concerns were taken into account or even acknowledged, production increased. In interviews, plant employees attributed the increase in their morale and productivity during the experiment to the feeling that someone cared about them, rather than to any changes in their physical environment. This phenomenon was called the "Hawthorne effect," after the name of the plant used for the study. This is an important point, providing managers with keys to affect productivity and morale. Managers do not have control over the physical environment in all instances, but they do have an opportunity to interact with their staff.

Contingency Theory

This theory was prompted by the notion that there is no one ideal method of management or organizational structure for all situations or types of business. The Contingency Theory holds that the organizational structure should be dependent on several factors, such as the function or purpose of the business, the environment in which the business operates (that is, either a stable or a changing environment), and the size or scope of the business. The type of management can also depend on the situation and the workers.

The situations that occur in a business setting require a variety of management styles. For example, many business decisions are best made by **consensus** (by a group). Sometimes the business or the situation is best served by a decision made by one person; for example, if the restaurant is on fire, does the manager wait for a consensus to determine whether everyone should leave, or does the manager simply evacuate the premises? Managers need to know their workers and the jobs they are to complete in order to decide how best to manage them. Remember, different situations require different types of management.

Managers and corporate planners can learn from all four schools of organizational theory. The physical environment and setup of the workplace must be examined, as recommended by Taylor's Scientific Management theory. In addition, the organizational structure must be designed for maximum effectiveness, as suggested by Fayol's Classical School. Managers must also be aware of the Hawthorne effect—that workers' outlook and feelings can significantly affect productivity. Finally, managers must realize that no single method of organization or type of management is best in all cases. They must learn which type of organizational structure and management style will work best to accomplish the job at hand.

CLASSIFICATION OF FOODSERVICE POSITIONS

The job titles in the foodservice industry may vary from operation to operation, but the responsibilities and roles are generally the same. The positions in foodservice operations are generally grouped into three categories: managers, production personnel, and service personnel (see Table 3.2).

Managers

The role of the manager is to ensure the smooth operation of the foodservice establishment. **Managers are responsible for making the decisions that steer the operation through the complex environment in which it competes.** They are also responsible for uniting the various departments, such as kitchen, dining room, and banquet service, toward the goal of the business. All areas must work together to achieve the objectives of the operation.

The position of manager is further broken down into three levels: top managers, middle managers, and supervisors. The responsibilities of managers at these levels vary, depending on the size of the operation. There are, however, some basic generalities that apply.

Top-Level Managers.   General manager, executive chef, and food and beverage director are examples of top management positions.

TABLE 3.2 Classification of Foodservice Positions

Managers	Top-Level Manager Middle-Level Manager Supervisor
Production Personnel	Kitchen Staff
Service Personnel	Dining Room Staff

Top managers are generally more concerned with and responsible for the long-term planning and goals of an operation than for day-to-day operations. They are responsible for monitoring trends in the industry and in the economy to determine the best course for the operation to follow. They delegate the day-to-day management of the operation to their middle-level managers, while still maintaining responsibility.

Middle-Level Managers.   Restaurant managers, bar managers, and catering managers are examples of middle managers in foodservice operations. Whereas top managers are more concerned with long-term planning, middle managers are responsible for the daily management of their operations. They are generally responsible for purchasing supplies, scheduling staff, and overseeing supervisors. Middle managers are also responsible for communication between top-level managers and supervisors.

Within the category of middle manager is the position of assistant manager. A person who attains the position of assistant manager has either worked his or her way up from supervisor or has been recruited out of college. Assistant managers work closely with managers, fill in when the managers are off duty, and generally are responsible for supervising supervisors.

Supervisors.   Line supervisor, lead cook, dining room supervisor, and lead waitperson are examples of supervisory positions. Generally, the first managerial position for a worker is that of a supervisor. Of the three levels of management, supervisors perform the most hands-on tasks, working directly alongside production or service personnel. They also have the added responsibility of keeping an eye on things and ensuring the quality of the food products and service. Supervisors provide valuable information to the upper levels of management.

Production Personnel

Employees who work in the "back of the house" (the kitchen) are classified as **production personnel**. They are the workers who are responsible for procuring, storing, and preparing the foods. The majority of their work is performed out of sight of the guests, although they too must keep the guests as their top priority. A production staff may have all three levels of managers, as mentioned earlier, depending on the size of the operation. See Table 3.3 for a description of production positions.

Service Personnel

The employees of a foodservice operation that spend the most time with the guests are the "front-of-the-house" staff, or **service personnel**. The extensive amount of customer contact associated with these positions makes them crucial to the success of the customer's dining experience. Table 3.4 gives a general description of the various service personnel positions.

SUMMARY

All four types of foodservice systems have advantages and disadvantages. Any of these systems, in whole or in part, can benefit an operation and the customers it serves. Many operations use a combination of the four, rather than any single system in its pure form. The development of these different systems shows that the foodservice industry is moving with the times and adapting to a changing environment.

A consideration of the various organizational approaches can provide insight into the evolution of thought in the realm of management. As organizational theory has evolved from Scientific Management, which viewed workers as simple tools of production, to the Contingency Theory, which recognizes that the best management style is dictated by many factors, the various forms of organization have developed.

The role of management is one of the most challenging that a worker will experience. A manager is only as good as the workers who work with and for him or her. A manager cannot supervise workers at all times, such as in all transactions with guests or during

TABLE 3.3 Food Production Personnel

Title	Description
Entry-Level Positions **(Require minimal or no prior training)**	
Kitchen Assistant (Prep Cook)	Assists cooks, chefs, and bakers in measuring, mixing, and preparing ingredients.
Mid-Level Positions **(Require minimum training and experience. On-the-job training is often provided.)**	
Cook	Prepares and places food on plates. In larger operations, may be responsible for specific foods, such as soups, vegetables, meats, or sauces.
Pastry Chef	Bakes cakes, cookies, pies, and other desserts, as well as bread, rolls, and quick breads.
Pantry Supervisor	Supervises salad or sandwich assistant. Should be able to create attractive food arrangements.
Upper-Level Positions **(Require both education and experience.** **Job descriptions may vary from one operation to another.)**	
Foodservice Manager	Responsible for profitability, efficiency, quality, and courtesy in all phases of the foodservice operation.
Assistant Manager	Performs certain supervisory duties under the manager's direction.
Food Production Manager (Sous Chef)	Responsible for all food preparation and supervision of the kitchen staff. Knowledgeable in food-preparation techniques, quality and sanitation standards, and cost-control methods.
Executive Chef	Responsible for quantity and quality of all food preparation for the entire operation, supervision of sous chefs and cooks, and menus and recipe development.

Source: Adapted from *The Choice of the Future: Careers in Foodservice,* The National Restaurant Association: The Educational Foundation, 1991.

preparation of all menu items. Therefore, a manager must be able to instill the standards of the operation in staff members so that they adhere to those standards without direct supervision. The better a manager can manage personnel, the more effective he or she will be. Managers can increase their effectiveness by understanding some of the basic principles of management.

The job titles of the positions in foodservice operations may vary from one operation to the next, but the responsibilities and roles are

TABLE 3.4 Food Service Personnel

Title	Description
	Entry-Level Positions **(Require minimal or no prior training)**
Bus Person	Clears and resets tables with fresh linen and silverware, refills water glasses, and assists food servers in serving and housekeeping chores in the dining room area.
Food Server	Takes customer's order and passes it to the kitchen along with special instructions, serves food and beverages, fills out guest checks, and sometimes takes payment. May be responsible for final assembly or preparation of some menu items; may be called on to offer menu suggestions.
Host/Hostess	Greets patrons, takes care of reservation lists, and shows guests to tables. Ensures that there is order and cleanliness in the dining area.
	Mid-Level Positions **(Require minimum training and experience. On-the-job training is often provided.)**
Dining Room Supervisor	Supervises all dining room staff and activities, including staff training, staff scheduling, and assignment of work stations. May also have the title of *wait captain* or *head server*.

Source: Adapted from *The Choice of the Future: Careers in Foodservice*, National Restaurant Association: The Educational Foundation, 1991.

generally consistent. The positions in foodservice operations are generally of three categories: managers, production personnel, and service personnel. Within each category are different levels of positions. Some positions require an educational background, and others simply require on-the-job training.

KEY TERMS

conventional system
traditional foodservice system
prepared in-house
"to order"
commissary
ready-food system

sous-vide
convenience food system
organization
labor-intensive
scientific management
classical school
human relations school

contingency theory
consensus
managers
top-level managers

middle-level managers
supervisors
production personnel
service personnel

DISCUSSION AND REVIEW QUESTIONS

1. Name two factors that have caused the development of new foodservice systems.

2. Which foodservice system is most similar to the method used by most families when cooking at home?

3. What causes the erratic patterns of foodservice activity? Which foodservice system is designed to compensate for such variation?

4. Name three benefits of using a *sous-vide* system. Can any restaurant process food using this method? Explain.

5. Develop a plan for a restaurant. What factors would you consider in deciding which type of foodservice system to use in your operation? Explain.

6. Complete the following table by listing the advantages and disadvantages of each of the four foodservice systems. Include the types of operations for which each system would be best suited.

	Type of Operation			
	Conventional	Commissary	Ready-Food	Convenience
Advantages:				
Disadvantages:				

7. Which two organizational theories discussed in the text stress the structure and the nature of the organization? What are the shortcomings of these two theories?

8. Why is the Contingency Theory the most modern of the theories discussed in the text? How was it developed?

9. How can a manager benefit from the use of the four schools of organizational theory? Tell how you would use an aspect of each in your role as manager of the restaurant developed in question 5.

10. What are the three levels of managers in a foodservice operation? In what ways are their responsibilities similar, and in what ways do they differ?

11. Why are service personnel so crucial to a guest's dining experience?

SUGGESTED READINGS

Axler, B. *Foodservice: A Managerial Approach* (New York: John Wiley & Sons, 1988).

Hackman, Lawler, and Porter. *Perspectives on Behavior in Organizations* (New York: McGraw-Hill, 1983).

Morgan, Jr., W. *Supervision and Management of Quantity Food Preparation* (Berkeley, CA: McCutchan Publishing, 1988).

Spears, M. *Foodservice Organizations*, 2nd ed. (New York: Macmillan, 1991).

CHAPTER FOUR

NUTRITION AND THE FOODSERVICE INDUSTRY

CHAPTER OBJECTIVES

Upon completion of this chapter, you should be able to:

- State the roles of fiber, fat, and cholesterol in the human diet.
- Name five factors that help people choose certain foods.
- State two roles that food plays in our society and culture.
- Discuss two factors that cause confusion about nutrition in America.
- Name the different responsibilities of the foodservice operator for both captive and free-to-choose patrons.
- Describe the different attitudes of customers toward nutrition when dining out.
- Discuss the characteristics and dining choices of the patrons in the three categories of nutritional concern.
- State some changes that foodservice operators are making to offer more acceptable choices to customers with nutritional concerns.
- Describe the response of table-service restaurants regarding the nutritional concerns of their customers.

A 1988 National Restaurant Association/Gallup survey asked consumers at table-service restaurants to rank various items according to their interests. Menu items for nutrition-conscious consumers were ranked second only to separate sections for smokers and nonsmokers. Of the customers surveyed, 74% selected the availability of nutritional choices as something they are interested in or very much interested in. Not all customers have jumped on the nutrition bandwagon, however, and some feel that this trend is still a fad.

Understanding their patrons' concerns, along with learning some basics of nutrition, can help foodservice operators to provide menu selections to satisfy the widest range of customers. In general, most operations do not need to overhaul their menus completely. Simply offering choices, adding or adapting a few items to appeal to health-conscious consumers, is sufficient.

Foodservice customers in general prefer a few nutritious choices on the menu. Yet, even though customers may understand the relationship between nutrition and wellness, taste is the primary criterion for food selection. The challenge for foodservice operators is to develop new recipes, or alter old favorites, to provide nutritious choices while making sure that the product tastes good.

NUTRITION FROM THE CUSTOMER'S POINT OF VIEW

A study of nutrition and the nutritional concerns of foodservice patrons is critical for foodservice managers. As the dining public becomes more knowledgeable about nutrition, and continues to be more concerned about health, it will become more demanding of nutritious food items when dining out. The better prepared managers are, the better they will be able to serve and satisfy their guests and help to ensure repeat business.

A National Restaurant Association (NRA) study discovered that customers can be grouped in three categories in regard to their attitudes toward nutrition when dining out. These range from the committed patron who will order only foods considered healthful, to the patron who is basically unconcerned with making healthful choices. The results of the study, combined with demographic information on each type of patron, provide valuable information to foodservice managers. The purpose of this chapter is to present an overview of nutrition and to discuss the importance of adapting menu items to provide customers with healthful food choices.

AN OVERVIEW OF NUTRITION

Nutrition is the study of foods and their relationship to health. Good nutrition requires the consumption of foods that are low in fat, high in fiber, and high in nutrients. An important point to remember is that good nutrition depends more on lifelong eating habits than on the sporadic consumption of a few items that are considered nutritious.

The general nutritional habits of Americans are generally considered to be poor. As a society, Americans eat too much fat and not enough fiber. In other words, many Americans eat too much of the things that are not nutritionally beneficial, and too little of the foods that are. The nutritional habits of the nation are improving, however, as people become more aware of the relationship between good eating habits and good health.

Food is the material that fuels the human body. The nutrients provided by the food we eat sustain our lives and give us the energy to perform our daily tasks.

The foods you eat provide your body with:

Water   To offer an environment for your cells to function and thrive;

Fuel   For energy, so that the body can do its work; and

Nutrients The "building blocks" that cells use to grow and maintain themselves.

A human body renews its internal and external structure continuously, each day building and replacing a small amount of its crucial components. A portion of the food a person eats becomes part of his or her internal structure. Remember the adage "You are what you eat"? The best food for a person to eat is that which supports the growth and maintenance of the internal components of the body while also cleansing and nourishing all parts of the body.

The components of food discussed in this chapter are carbohydrates, dietary fibers, fat, cholesterol, proteins, and other nutrients. A brief description of each is given to indicate their importance in the diet (see also Table 4.1). For a more detailed discussion of the topic, refer to a textbook on nutrition (see the "Suggested Reading" section at the end of this chapter).

Carbohydrates contain the sun's energy captured in a form that organisms can use. They are found primarily in foods derived from plants, with the exception of milk, the only animal-based food that contains adequate levels of carbohydrates. Carbohydrates are primarily sugars and starches.

Dietary fibers are also carbohydrates, but these substances cannot be broken down by the human digestive tract. They are beneficial because they promote a feeling of fullness and help to prevent intestinal problems by keeping things moving through the intestines. Dietary fiber is found in fruits, vegetables, legumes, and whole grains.

Fat is generally recognized as the part of the diet that has the potential for harm to a person's health. Yet what most people do not understand is that fat is also essential for the function of the body. Fat is the body's main form of storage for the energy provided by food in excess of that which the body needs for immediate use. However, when a person consumes more fat than is needed, the excess fat accumulates in the body and causes health problems. Fats occur in most foods in many different forms and types, some better for the human body than others. Those that are the worst for the body are the fats derived from animal sources, and those that are generally best are the fats from plant sources.

Cholesterol is a soft, waxy substance manufactured in the body for a number of purposes; it is also found in animal products. Cholesterol is another substance essential to the functioning of the human body. However, the body produces all of the cholesterol it needs; therefore, when a person's diet contains food high in cholesterol, a problem arises. Cholesterol builds up and lines the arteries and can restrict the flow of blood to the heart. Elevated levels of cholesterol are the primary cause of cardiovascular disease, the number one cause of death in this country. The level of cholesterol in a person's blood is considered to be a predictor of his or her chances of having a heart attack. A high level of cholesterol reflects increased blockage of the crucial blood flow to the heart. Among other foods, red meats are generally high in cholesterol.

Proteins serve several purposes in the human system. Their primary function is to help with the growth and maintenance of the cells that make up the body. They provide a catalyst for the chemical reactions that are necessary for life. Meats, fish, poultry, dairy products, and soybeans are all good sources of proteins.

Nutrients include the trace elements found in food that provide valuable components in growth and development. The body produces some of these itself, while others must be obtained by eating food rich in specific nutrients. Calcium, iron, zinc, and vitamins A, B, E, and K are nutrients that the body needs to function properly. The best way for a person to ensure that he or she gets all of these nutrients is to eat a balanced diet. The problem is that some nutrients are lost in the processing and cooking of foods. Fresh fruits and vegetables, grains, and legumes are rich sources of these nutrients.

TABLE 4.1 The Primary Components of Food

Component	Examples
Carbohydrates	Sugars, starches
Dietary Fiber	Fruits, vegetables, whole grains
Fat	Animal fat, vegetable oils
Cholesterol	Based in animal fat
Protein	Meat, fish, poultry, soybeans
Other Nutrients	Vitamins, minerals

HOW PEOPLE CHOOSE THE FOOD THEY EAT

How do people choose the food they will eat? Do they choose foods that will best meet their bodies' needs? Do they choose foods they like, or do they choose foods that are most convenient for them? It is important for foodservice operators to understand how their customers choose what to eat so that they can design their menus accordingly. A majority of people choose to eat food they like rather than food that is good for them. The challenge for the foodservice operator is to provide both good-tasting and healthful food.

A survey discovered several factors that people cite to explain their choices of food:

Personal preference: They like the item.

Habit: They are familiar with the item and eat it often.

Ethnic tradition: The food is part of their ethnic heritage.

Availability: Food choices were limited.

Convenience: It was convenient for them when they were hungry.

Economy: It was affordable.

Nutritional value: They thought the food item was good for them.

The survey showed that people generally base their food choices on behavioral or social factors, rather than on the importance of what they eat to their physical well-being. The factors listed here are beneficial both to the study of nutritional concerns and to the entire foodservice industry. A foodservice operator can better plan an operation when the motivations for food choices are understood.

CHANGES IN FOOD CONSUMPTION PATTERNS

There is evidence that the eating habits of Americans are improving, as provided by data on consumption of key food items. The overall national consumption of red meat (beef, pork, and lamb) has been steadily decreasing over the last five years; interestingly enough, the consumption of red meat has remained fairly constant in foodservice operations. The consumption of poultry and fish has increased because of the low fat content of these foods. Changes in consumption of dairy products, reflect a shift toward low-fat milk and cheeses, is also indicative of a move toward a healthier diet. Egg consumption has been dropping steadily since eggs were shown to be high in cholesterol. New methods for measuring the cholesterol in eggs have revealed, however, that eggs are not as high in cholesterol as previously thought. Yet unfortunately for the egg industry, the updated test results have not been able to reduce significantly the decline in egg consumption. Alcohol consumption has also been changing—hard liquor consumption has been steadily declining since 1969 because of the fact that it has very little nutritional value, while beer and wine consumption has risen slightly.

THE ROLE OF FOOD IN OUR SOCIETY

Food nourishes the body, and it nourishes the mind and spirit as well. Deeply rooted in our culture, food plays an important role in our social functions. Celebrations, weddings, birthdays, most holidays, and other celebrations usually center on a meal. Business meetings, sales meetings, and conventions are typically preceded or followed by a meal. Providing food is considered a way of displaying affection, as evidenced by inviting others to share a meal or giving food as a gift. Certain foods can be fun, or are traditional for certain rituals or holidays in our culture, such as hot dogs at a barbecue or turkey at Thanksgiving dinner.

The foodservice industry is repeatedly blamed for the poor nutritional habits of the nation. On average, Americans eat one of every four meals away from home. Restaurant menus are generally loaded with deep-fried items, foods covered with rich sauces, or highly processed food items, most of which are lacking sufficients nutrients. Foodservice operators respond that they serve what the customer wants, that if people would stop ordering deep-fried foods, they would stop serving such items.

PUBLIC AWARENESS OF NUTRITION

The American public's awareness of good nutritional habits has increased with the "health craze" that has swept the nation. At first considered a fad, the combination of improved eating habits and increased physical fitness is now considered to be a permanent part of Americans' lives.

The problem with the increased attention to improved nutrition is the general state of confusion fueled by media sensationalism. To further complicate the matter, Americans seem to be more interested in quick fixes for problems, rather than long-term remedies. Generally, people want a pill or a miracle food they can consume that will reverse a lifetime of improper eating. The media fuels the problem by giving widespread coverage to innovations that may not have been properly researched and can turn out to be ineffective.

Nutritional quackery, or fraudulent nutritional information, stands in contrast to the reliable nutritional information that is widely available. Unfortunately, the field of nutrition is full of fraud, with dishonest producers and salespeople making unfounded claims about the products they sell. The lack of standard definitions for such terms as "light" and "natural" opens the door for unscrupulous food producers to abuse these nonstandard terms for their own gain. This further confuses consumers and makes it harder for the average American to choose wisely.

An example of the results of highly publicized but improper research is the nationwide reaction to the announcement of the effect of oat bran on the lowering of cholesterol. The study was initially funded by a company that processes oat bran and was performed on only a relatively small number of people. The results were released at a press conference, which dramatically changed the national perception of nutrition. In a short period of time, the price of oat bran skyrocketed and many people pursued this miracle remedy to their cholesterol problems. Food companies were able to bring all types of new products that contained oat bran, from chips to donuts, to the consumer market in record time. Then another study came out refuting the results of the first study. The second study found that oat bran was no more effective in lowering cholesterol than any other type of bran. This new study was plastered over all the television and newspapers. The coverage by the popular press distorted the results of the second study, indicating that oat bran was not effective in lowering cholesterol, rather than reporting the truth of the results, which provided that oat bran was no more—or less—effective than any other type of bran product. The most significant result of these conflicting reports was to confuse consumers.

THE FOODSERVICE INDUSTRY'S RESPONSIBILITY

Because, on average, Americans consume one of every four meals away from home, what exactly is the foodservice industry's responsibility toward the nutrition of the nation? Upon closer examination, those eating meals away from home can be grouped in two major categories—captive audiences and those who are free to choose.

CAPTIVE AUDIENCE

Captive audience consumers are those who have little choice but to dine at a certain establishment—those in prisons, school dormitories, hospitals and other institutions—because their meals are either prepaid or paid as part of a room fee or other charge. In these cases, the responsibility to provide well-balanced meal choices is especially critical for the foodservice operator. Theoretically, the foodservice operator is providing the only food available to customers in this segment of the industry and, therefore, must plan carefully to provide healthful choices.

Foodservice operators who serve a captive audience usually employ a **nutritionist** or registered dietitian to assist them with their meal planning. Nutritionists and registered dietitians are professionals who have studied nutrition and who have been certified for their expertise in the relationship between proper diet and health. Because different groups have different nutritional needs (for example, those in grade school, in high school, or in penal institutions), the nutritional requirements of the group being served are evaluated and taken into account in menu development. Planners take care to ensure that the patrons are offered correct proportions of the various necessary nutrients.

The foodservice operator's responsibility stops with the providing of healthful choices for meals. He or she cannot be held responsible for making sure the patron chooses wisely. College dormitory cafeterias have changed quite a bit in recent years in response to students' desire for more healthful food choices. Many college foodservice facilities offer salad bars, nonmeat main dishes, and other food items that are considered to be healthful. The decision to choose between a serving of steamed vegetables and brown rice and a meal of hamburger and french fries lies with the patron, not with the operator, of the foodservice.

FREE TO CHOOSE

Customers who have a choice as to where they dine are included in the **free-to-choose** or pay-per-meal category. They have the option to dine at a fast-food restaurant, a fine-dining restaurant, a snack bar, or the deli of a grocery store. Because the foodservice operator is not the sole provider of the food for this customer, the responsibility of the operator to provide nutritional food items is much less than that at an operator serving a captive audience. It is the customer who can decide whether he or she is going to eat healthful food or junk food, depending on the foodservice operation he or she chooses. Of course, customers need basic education in nutrition to assist them in their menu selections. The key is *choice*.

Operators serving these customers are not totally free from having to provide healthful food choices. Customers who dine in their establishments may very well want healthful items for their meals. The competitive nature of the business environment in which they operate forces operators in this segment of the industry to be much more receptive to the wants and needs of their customers.

THE CUSTOMER'S ATTITUDE TOWARD NUTRITION WHEN DINING OUT

In 1986, 1989, and again in 1992, the NRA conducted a nationwide study to assess consumers' awareness and attitudes about nutrition. It also sought to determine how those attitudes affect consumers' choices of foods when dining away from home. The major finding of the survey is that foodservice consumers can be grouped in three distinct categories—unconcerned, vacillating, and committed. The information provided by the survey, summarized in Table 4.2, gives valuable insight into how customers view nutrition when dining out

TABLE 4.2 Characteristics of Consumer Groups Regarding Health and Nutrition, 1989

Characteristics	Unconcerned	Committed	Vacillating
Percentage of U.S. Adult Population	32%	37%	31%
Percentage of Total Eating-Out Occasions	39%	32%	29%
Demographic Characteristics	Men, 18 to 34 years old, average income, live in southern states	Women, 35 to 54, above average income, married, live in metropolitan areas	Women, 45 and older, below average income, live in Northeastern States below average income
Behavioral Characteristics	Patronizes fast-food restaurants, do not diet or exercise, drink alcoholic beverages, do not restrict fat or calories	Patronize moderately priced and fine-dining restaurants, diet and exercise, restrict intake of fat and calories	Patronize both fine-dining and fast food, diet to control cholesterol and high blood pressure
Food Likely to Order When Dining Out	Steak or roast beef, fried chicken or fish, regular soft drinks, rich desserts	Broiled or baked seafood, poultry without skin, frozen yogurt	Lean meats, steaks, fried foods, regular soft drinks

Source: National Restaurant Association, *Nutrition and Restaurants: A Consumer Perspective,* 1993.

and how foodservice operators can develop their menus (see also Figure 4.1).

THE UNCONCERNED PATRON

Unconcerned patrons are usually not interested in nutrition or health—they generally are considered to be meat-and-potato eaters. They are more likely to eat at fast-food restaurants and are generally likely to be males between 18 and 34 years old. These patrons are good restaurant customers, representing 32% of the total meals eaten out, although they make up only about 32% of the U.S. adult population. They are not likely to order low-fat, low-calorie menu items, nor are they very big salad eaters. They tend to order red meat, fried food, regular soft drinks, and rich desserts.

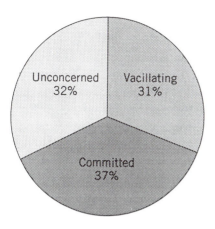

FIGURE 4.1 The dining population: concern about nutrition.

THE VACILLATING PATRON

Vacillating patrons are concerned about nutrition and eating healthful foods, but they are driven by taste and occasion when eating away from home. Their menu choices are not consistent—they are just as likely to order rich items and red meat as they are to order lean items such as broiled fish and salads. They may order decaffeinated coffee with a sugar substitute to accompany a rich slice of cheesecake for dessert. Vacillating patrons tend to be older females and below-average income earners. These customers generally eat differently at home than they do when dining out.

THE COMMITTED PATRON

Committed patrons firmly believe that there is a relationship between their health and their nutrition and diet. They carry their commitment to nutrition with them when dining out. They are likely to order the items that the unconcerned patron avoids. These concerned patrons typically order broiled or baked fish or poultry and whole grain muffins or rolls, expect vegetables to be served, and like fresh fruit for dessert. Those in this group comprise above-average-income females who generally live in metropolitan areas. They firmly believe that restaurants should offer healthful choices at all meals. They generally dine at table-service restaurants, but dine out less than the average.

INTERPRETING THE DATA ON PATRONS' VIEWS OF NUTRITION

Most people dine in groups of two or more. When choosing a place to dine, most groups attempt to find foodservice facilities that will provide choices agreeable for all members of the group. If a restaurant does not serve any of the "healthful" food that is preferred by committed patrons, it will not only lose the 32% of the population that prefers that type of food, but it will also lose the business of the group with which the committed patron dines. Similarly, if an operation serves only healthful food, it will lose more than just the unconcerned patron; it will also lose the group with which the unconcerned patron dines.

The message to foodservice operators is that to be successful and to appeal to a wide range of customers, they must offer menu choices for all types of consumers. Except in extreme cases, it is not advisable for an operation to serve food strictly for committed patrons or to serve food only for unconcerned patrons. Some of the recent changes in the menu offerings of fast-food chain restaurants indicate a move to appeal to a broader range of clientele. Fast-food operations that previously served only hamburgers and fries are now serving salads, broiled chicken, low-fat hamburgers, and low-fat frozen yogurt. To supplement these new, more healthful choices, fast-food operations are also providing nutritional information on the products they serve so that consumers can make more informed choices about their meals.

THE FOODSERVICE INDUSTRY'S RESPONSE TO THE NUTRITIONAL CONCERNS OF THEIR GUESTS

The foodservice industry has been adapting to satisfy the demand for more healthful food choices. Menus that include raw vegetable appetizers, broiled or baked skinless poultry or fish, and fresh fruit or low-fat yogurt desserts are appearing in a variety of foodservice establishments. The following paragraphs discuss the changes being made in the foodservice industry.

CHAIN RESTAURANTS

The NRA conducted a survey in 1991 to assess several aspects of how nutritional concerns are met in the operation of chain restaurants. The results of this survey are shown in Table 4.3. It found that the majority of the chain restaurants surveyed provided a

TABLE 4.3 Survey Results of Chain Restaurants'
(350+ Operations) Nutritional
Offerings, 1990

Item	Percentage of Chain Operators Serving
Decaffeinated coffee	89
Margarine	67
Entree salads and salad bars	78
Decaffeinated soda	41
Reduced- or low-calorie salad dressing	74
Grilled chicken sandwich	59
Fresh fruit	48
Low-fat frozen yogurt	33

Source: National Restaurant Association, *Nutrition and Restaurants: A Consumer Perspective*, 1993.

number of menu items to allow their guests to have more healthful choices. Decaffeinated coffee, low-fat milk, entree salads, fruit juices, and reduced-calorie salad dressings were among the offerings.

TABLE-SERVICE RESTAURANTS

Diners at table-service restaurants generally have a high degree of interest in nutritional menu items. Operators of this type of foodservice are providing menu offerings to satisfy the more health-conscious consumer. Besides offering healthful items, they also alter or adapt menu items to satisfy nutritional concerns. See the results of a 1988 NRA survey in Table 4.4.

Preparation methods are generally not difficult to alter. Table-service operations, most of which prepare their menu items to order, should be able to handle simple requests to help satisfy customers' nutritional concerns. The advantage of these methods is that they allow an operation to expand its number of healthful food items and, therefore, increase the number of people it satisfies, without adding to its inventory and without any major menu or equipment changes. Thus, operations can increase customer satisfaction without any additional expense and only a minimum of additional labor. An operation that is concerned with attracting and satisfying the

TABLE 4.4 Changes That Restaurants Make in Altering Preparation Methods Upon Request, by Average Dinner Check Size per Person

	$8.00	$8.00–$14.99	$15.00–$24.99	$25.00+
Percentage of operations that will alter their preparation methods upon request	85	93	92	98
Types of Changes by Those That Will Alter Methods				
Serve sauce on the side	99	100	100	98
Serve salad dressing on the side	99	100	100	99
Cook without salt	85	89	99	99
Broil or bake, rather than fry	78	95	98	99
Prepare food using vegetable oil/margarine rather than butter or shortening	89	95	98	92
Skin chicken before preparing	81	88	94	10
All the above changes	50	70	73	87

Source: National Restaurant Association, *Table Service Trends: 1992.*

health-conscious patron can print on the menu the preparation method it is willing to alter so that customers can be aware of those variations without having to ask.

Table-service operations can also offer healthful items to supplement their regular menu. This will encourage the concerned patron and expand the range of offerings for the vacillating patron, without scaring away the unconcerned patron.

Further results of the 1992 NRA survey, summarized in Table 4.5, clearly show that a majority of foodservice operators in the table-service segment of the industry are providing healthful choices to supplement their meal offerings. Note that the items listed in the table are not difficult to obtain or stock for the majority of table-service restaurants, or for any type of operation.

SUMMARY

Foodservice operators cannot expect to survive today without considering the nutritional concerns of their customers. This is not to say that restaurants should go to such drastic lengths as to remove all red meat from the menu—the point is that restaurants must offer healthful choices for their customers. Foodservice operators must

TABLE 4.5 Percentage of Restaurants Offering "Healthful Items" to Supplement Their Menu Offerings by Dinner Check Size

	$8.00	$8.00–$14.99	$15.00–$24.99	$25.00+
Sugar substitute	99	99	96	94
Diet beverages	100	100	100	99
Caffeine-free beverages	88	91	84	92
Fresh fruit for dessert	56	62	62	91
Whole grain bread, rolls, or crackers	81	74	68	69
Margarine	87	88	83	77
Low-fat or skim milk	59	55	49	57
Reduced-calorie salad dressing	70	65	40	38
Salt substitute	30	18	21	11
All of the above	5	7	3	3
7–8 of items above	43	32	25	33
4–6 of items above	44	54	59	58

Source: National Restaurant Association, *Table Service Trends: 1992.*

understand the different views on nutrition which are generalized into three groups of patrons—committed, unconcerned, and vacillating—and successful operators will provide menu items accordingly.

KEY TERMS

nutrition	nutritional quackery
carbohydrates	captive audience
dietary fibers	nutritionist
fat	free-to-choose
cholesterol	unconcerned patron
proteins	vacillating patron
nutrients	committed patron

DISCUSSION AND REVIEW QUESTIONS

1. Explain why it is in the best interest of a foodservice operation to offer healthful menu selections for its guests.

2. Are all fats bad for humans? Are some beneficial? Explain.

3. What is the primary factor that motivates most peoples'

choices of the food they eat? How does this factor pose a challenge to foodservice operators?

4. Name two of the changes in food consumption patterns in this country, and discuss how these changes may affect the foodservice industry.

5. The foodservice industry has come under fire for the general decline in the nutritional standards of the country. Is that blame warranted? Explain.

6. How does the obligation of providing nutritional meal choices differ between foodservice operators that serve a captive audience and those that serve customers who are free to choose?

7. You are employed as the foodservice manager of a moderately priced table-service restaurant. The chef feels that the concern with nutrition is just another passing fad and does not see the need to provide healthful choices on the menu. How would you explain to him or her the potential consequences of this position? Include a discussion of the three types of customers, and tell why you may be interested in attracting more than the group of customers who are members of the "committed group"?

8. List four ways in which a foodservice operation can make its menu more healthful without totally overhauling it.

SUGGESTED READINGS

Drummond, K. *Nutrition for Foodservice Professionals* (New York: Van Nostrand Reinhold, 1994).

Hamilton, Whitney, and Sizer. *Nutrition Concepts and Controversies,* 5th ed. (St. Paul, MN: West, 1991).

Hodges, C. *Culinary Nutrition for Foodservice Professionals* (New York: Van Nostrand Reinhold, 1994).

Poleman, Charlotte. *Nutrition Essentials and Diet Therapy,* 6th ed. (Philadelphia: Saunders, 1991).

CHAPTER FIVE

SANITATION IN FOODSERVICE OPERATIONS

Most people generally agree that a foodservice operation must be kept clean. This chapter discusses why it is important to keep a foodservice operation both clean and sanitary. Legal, moral, and monetary reasons make it imperative for foodservice operations to keep all aspects of their workplaces clean.

SOME PRELIMINARY DEFINITIONS

To begin a discussion of sanitation, some basic terms must be defined:

Clean	Free from visible soil.
Contamination	The unintended presence of harmful substances or microorganisms in food or beverages.
Cross-contamination	Allowing harmful substances to come in contact with new products.
Food-borne illness	A disease that is carried or passed to human beings by food.
Food-borne illness outbreak	A reported incident of two or more people becoming ill from a common food, which is confirmed through laboratory analysis as the source of the illness.
Sanitarian	A health inspector or official trained in sanitation principles and methods; a representative of the state or local board of health.
Sanitary	Free of disease-causing bacteria.
Sanitation	The creation and maintenance of healthful conditions. Wholesome food handled and prepared in such a way that it is not contaminated with disease-causing bacteria.
Spoilage	Damage to the edible quality of food.

The difference between **clean** and **sanitary** is illustrated by the following: It is possible for an item, for example, a dish or serving piece, to be clean but not sanitary. When you examine a plate and do not see any visible soil, you may describe it as clean. Yet because bacteria are too small to be visible to the human eye, the plate could be covered with harmful bacteria and, therefore, not sanitary.

Future managers of foodservice establishments must be made aware of some basic concepts and principles. Because some food hazards are caused by bacteria, it is important to know what bacteria need in order to live and the various ways in which they are spread.

SANITATION FROM THE CUSTOMER'S POINT OF VIEW

Customers rate and judge the cleanliness of a foodservice operation in all phases of their dining experience. Both consciously and unconsciously, they use their sense of smell to determine whether an operation is acceptable or objectionable. They scan the appearance of the exterior and interior of the facility to determine the general level of maintenance and upkeep. They judge the staff of the operation by their personal hygiene and the cleanliness of their uniforms. Customers combine their judgments in all of these areas to arrive at an overall decision on the cleanliness and sanitation of the operation.

The problem or benefit, depending on one's perspective, is that customers are permitted to see only a portion of the foodservice operation and staff. Customers are allowed to see the front-of-the-house staff—host/hostess, servers, and bus people—but usually not the food-preparation areas and staff. Generally, customers will make a judgment on the cleanliness and personal hygiene of the kitchen staff based on the staff they *are* allowed to see. The saying "You can judge the cleanliness of a restaurant kitchen by the cleanliness of its restrooms" is based on the same premise. If management has allowed a visible area of the operation to become dirty or messy, what might they allow in the areas of the operation customers cannot see?

The National Restaurant Association (NRA) conducted a survey in 1991 to determine how important to customers is the cleanliness of a foodservice operation. The NRA chose customers of three types of foodservice operations—quick service, moderate service, and full service—and asked them to rate the importance of a group of characteristics in choosing a place to dine. Customers rated cleanliness as the most important in two types of restaurants, and second to food quality and preparation in the third. If the goal of an operation is to meet or exceed the expectations of its customers in order to succeed, then cleanliness is essential, as indicated by the results of this survey.

Food naturally contains safe levels of both harmful and safe bacteria. The mishandling of food—leaving it at room temperature for too long—will allow bacteria to grow to sufficient numbers to cause food poisoning or food-borne illness. Although food poisoning and food-borne illnesses are seldom fatal, except for the very young and the very old, they do cause severe symptoms. A person suffering from a food-borne illness experiences almost unbearable pain and discomfort. This is an illness that is generally avoidable, so most people are wary of places to eat that appear dirty or unsanitary. Cleanliness, or lack of it, becomes an important advertisment, either positive or negative, for the establishment.

It is interesting to note that when a person experiences the symptoms of food poisoning—nausea, vomiting, and abdominal cramps—he or she often places the blame on a meal eaten at a foodservice establishment. Perhaps the reason for this almost-certain blaming of a commercial foodservice establishment is that people may not want to take personal responsibility. It is understandable that most food-borne illness outbreaks at restaurants go unreported. After dining, most restaurant customers leave the establishment and go in different directions. When they experience symptoms, they generally feel they have come down with a slight flu or a "bug." This is not true in foodservice operations that provide food for captive audiences, such as school cafeterias, cruise ships, prisons, or banquet and catering operations that serve people who have a relationship to one another. When these customers begin to experience symptoms, they are able to discuss their medical problems with one another and trace them to a particular meal or function they had attended.

THE IMPORTANCE OF SANITATION TO A FOODSERVICE OPERATION

The reasons that sanitation is important to a foodservice operation fall into three categories: legal, monetary, and moral (see Figure 5.1).

Three Reasons Why Sanitation Is Important to Foodservice Operations

Legal
Monetary
Moral

FIGURE 5.1 Reasons for the importance of sanitation in foodservice operations.

LEGAL REASONS

A foodservice operation does not have a choice as to whether or not it wants to keep its place clean—all are obligated by law to do so. The various levels of government agencies set regulations and standards for foodservice operators to follow, and the responsibility lies with the foodservice operator to abide by those laws and regulations. The government has neither the time nor the resources to monitor foodservice operators on a daily basis; it relies on the operators to follow the laws that have been established.

There are three levels of government agencies that have jurisdiction over foodservice operations: federal, state, and local.

Federal

The federal government has little to do with the daily operation of foodservice operators. The role of the federal government is to protect the quality of the food products the operator purchases. The three key federal agencies that have the greatest impact are the FDA, the USDA, and the CDC.

The **FDA** (Food and Drug Administration) is responsible for developing ordinances and regulations for state and local health departments. The model ordinances developed by the FDA become the basis for the state and local regulations and codes. State and local health departments have the option to use the FDA codes as written or to adapt them, as long as the changes are at least as strict as the FDA-proposed codes. The FDA also regulates the inspection of food-processing plants. Its role is to ensure that processing plants adhere to federal standards of wholesomeness and purity.

The **USDA,** (U.S. Department of Agriculture) inspects and grades meat, poultry, dairy products, eggs, and produce that is shipped across state lines (see Figure 5.2). The grading of products, which is optional, involves checking for quality and yield and does not reflect the sanitation or cleanliness of a product, whereas inspection is mandatory and is performed to ensure safety and wholesomeness. **Wholesomeness,** according to the USDA, characterizes a product that is free from disease and was processed under sanitary conditions.

The **CDC** (U.S. Centers for Disease Control) investigates and compiles data on food-borne illness outbreaks and studies the causes and controls of disease.

Although the regulatory agencies of the federal government have little to do with the daily operations of the average foodservice

Inspection Stamps

Grade   Food Quality Stamp

FIGURE 5.2 Inspection of grade stamps. (*Source:* U.S. Department of Agriculture)

operation, they do regulate the food used in foodservice operations. The regulation of daily operations and inspection of commercial foodservice operations that just produce food to be served on the premises is left to state or local health department officials.

ℰ State and Local

State and local health departments have the greatest influence on the daily operation of commercial foodservice operations. Jurisdiction depends on the size of the community. Larger cities generally have their own boards of health, whereas smaller communities fall under the jurisdiction of state or county boards of health. Regulations may vary considerably from state to state and community to community. The only central measure is that state and local regulations must be at least as stringent as the federal model ordinance. A board of health has the option to pass and enforce an ordinance that is stricter than the **FDA Model Sanitation Ordinance** if it sees fit.

The local board of health determines the regulations and codes for the operations that do business in its jurisdiction. It performs inspections of all foodservice operations on a periodic basis and is available for training and consultation on sanitation problems. In some areas plans for foodservice operations must be preapproved prior to construction. Some foodservice workers may be required to obtain a food-handlers card prior to working in a preparation area.

◦ Foodservice Managers and the Board of Health

The importance of foodservice managers working with (not against) a board of health cannot be overemphasized. Foodservice managers must be aware of the board of health regulations in their area. The FDA recommends that all operations be inspected at least once every six months. Foodservice operators should view the inspectors as allies in the job of maintaining a clean and sanitary operation—they are not the enemy. The goal of the foodservice operator should be to maintain a safe place to prepare and serve food, and the goal of inspectors (also known as **sanitarians**) is the same—to ensure that the operation is maintaining a safe environment. The frequency of the inspections depends on a number of factors: a complaint or group of complaints, the results of past inspections, and the workload of the department inspectors. The job of the foodservice manager is to ensure that the operation is kept clean daily, and the role of the inspector is to supplement, not replace, the inspections of the manager.

Inspectors have the right upon display of proper identification, to enter and inspect any commercial foodservice operation during business hours. Much valuable information can be learned from an inspector during an inspection. Remember, an inspector's aim is to assist the operation and the manager in maintaining a safe environment.

The following suggestions, derived from applied foodservice sanitation procedures, can enable a manager to get the most out of a sanitation inspection:

1. *Ask for identification.* Make sure that the person arriving for inspection has proper identification and is authorized to inspect the operation. Ask the purpose of the inspection; be sure to find out whether it is a routine inspection, the result of a customer complaint, or initiated for another reason.

2. *Cooperate.* Remember that sanitarians are also humans and treat them with courtesy. A display of evasiveness, resentment, or lack of cooperation can be interpreted as an indication that the establishment is trying to hide something. Keep in mind

that an inspector has a right by law to inspect the operation—the way the inspector is treated can have a definite effect on the outcome of the inspection. A manager should answer the inspector's questions and instruct employees to do the same, being truthful to the best of their knowledge. A manager should offer to accompany the sanitarian on the inspection; this shows that the manager is concerned and allows him or her to learn from the sanitation professional.

3. *Take notes.* Make notes of an inspector's comments as you accompany him or her on the inspection, remedying any problems that can be corrected immediately. Such involvement displays your concern and helps you to better maintain the operation.

4. *Keep your relationship professional.* Do not offer anything to an inspector that may be misinterpreted as a bribe.

5. *Be ready to provide records.* The board of health has the right to review various records of an operation. Be aware of the records that are required by regulation in the area of operation for the restaurant and have them ready just in case an inspector requests them.

6. *Discuss any violations with the sanitarian.* Sanitarians visit foodservice operations on a daily basis and are well versed in the regulations of the area. They are therefore a valuable source of information and advice on improving the sanitation of an operation. Managers should make sure they understand the nature of any violations that a sanitarian finds so that they know how to correct them and avoid such problems in the future.

7. *Follow up.* A manager should carefully go over his or her copy of an inspection report, then take it along while going through the facility. The next step is to correct the problems. Different violations require different follow-up action by the sanitarian. Some minor violations can be remedied quickly and easily and require no further action; other more serious infractions require a follow-up visit by the sanitarian. It is the responsibility of the foodservice manager to make sure all violations are corrected promptly.

Figure 5.3 is an example of an FDA Foodservice Establishment Inspection Report form. Notice that the various items on the form

DEPARTMENT OF HEALTH, EDUCATION AND WELFARE
PUBLIC HEALTH SERVICE—FOOD AND DRUG ADMINISTRATION

FOOD SERVICE ESTABLISHMENT INSPECTION REPORT

PURPOSE
Regular.... 29–1
Follow-up 2
Complaint 3
Investigation...4
Other..........5

Based on an inspection this day, the items circled below identify the violations in operations or facilities which must be corrected by the next routine inspection or such shorter period of time as may be specified in writing by the regulatory authority. Failure to comply with any time limits for corrections specified in this notice may result in cessation of your Food Service operations.

OWNER NAME

ESTABLISHMENT NAME

ADDRESS

ZIP CODE

EST. I.D. (1–10) | COUNTY | DIST. | EST. NO. | CENSUS TRACT 11–13 | SANIT. CODE 14–16 | 17–22 | YR. | MO. | DAY | TRAVEL TIME 23–25 | INSPEC. TIME 26–28

ITEM NO.		WT.	COL.
FOOD			
*01	Source; sound condition, no spoilage	5	30
02	Original container; properly labeled	1	31
FOOD PROTECTION			
*03	Potentially hazardous food meets temperature requirements during storage, preparation, display service, transportation	5	32
*04	Facilities to maintain product temperature	4	33
05	Thermometers provided and conspicuous	1	34
06	Potentially hazardous food properly thawed	2	35
*07	Unwrapped and potentially hazardous food not re-served	4	36
08	Food protection during storage, preparation, display, service, transportation	2	37
09	Handling of food (ice) minimized	2	38
10	In use, food (ice) dispensing utensils properly stored	1	39
PERSONNEL			
*11	Personnel with infections restricted	5	40
*12	Hands washed and clean, good hygienic practices	5	41
13	Clean clothes, hair restraints	1	42
FOOD EQUIPMENT & UTENSILS			
14	Food (ice) contact surfaces: designed, constructed, maintained, installed, located	2	43
15	Non-food contact surfaces: designed, constructed, maintained, installed, located	1	44
16	Dishwashing facilities: designed, constructed, maintained, installed, located, operated	2	45
17	Accurate thermometers, chemical test kits provided, gauge cock (¼" IPS valve)	1	46
18	Pre-flushed, scraped, soaked	1	47
19	Wash, rinse water: clean, proper temperature	2	48
*20	Sanitization rinse: clean, temperature, concentration, exposure time; equipment, utensils sanitized	4	49
21	Wiping cloths: clean, use restricted	1	50
22	Food-Contact surfaces of equipment and utensils clean, free of abrasives, detergents	2	51
23	Non-food contact surfaces of equipment and utensils clean	1	52
24	Storage, handling of clean equipment/utensils	1	53
25	Single-service articles, storage, dispensing	1	54
26	No re-use of single service articles	2	55

ITEM NO.		WT.	COL.
WATER			
*27	Water source, safe: hot & cold under pressure	5	56
SEWAGE			
*28	Sewage and waste water disposal	4	57
PLUMBING			
29	Installed, maintained	1	58
*30	Cross connection, back siphonage, backflow	5	59
TOILET & HANDWASHING FACILITIES			
*31	Number, convenient, accessible, designed, installed	4	60
32	Toilet rooms enclosed, self-closing doors; fixtures, good repair, clean: hand cleanser, sanitary towels/hand-drying devices provided, proper waste receptacles	2	61
GARBAGE & REFUSE DISPOSAL			
33	Containers or receptacles, covered: adequate number insect/rodent proof, frequency, clean	2	62
34	Outside storage area enclosures properly constructed, clean; controlled incineration	1	63
INSECT, RODENT, ANIMAL CONTROL			
*35	Presence of insects/rodents—outer openings protected, no birds, turtles, other animals	4	64
FLOORS, WALLS & CEILINGS			
36	Floors, constructed, drained, clean, good repair, covering installation, dustless cleaning methods	1	65
37	Walls, ceiling, attached equipment: constructed, good repair, clean, surfaces, dustless cleaning methods	1	66
LIGHTING			
38	Lighting provided as required, fixtures shielded	1	67
VENTILATION			
39	Rooms and Equipment—vented as required	1	68
DRESSING ROOMS			
40	Rooms, area, lockers provided, located, used	1	69
OTHER OPERATIONS			
*41	Toxic items properly stored, labeled, used	5	70
42	Premises maintained free of litter, unnecessary articles, cleaning maintenance equipment properly stored. Authorized personnel	1	71
43	Complete separation from living/sleeping quarters. Laundry	1	72
44	Clean, soiled linen properly stored	1	73

FOLLOW-UP
Yes 74–1
No..........2

RATING SCORE 75–77
100 less weight of
items violated ➝

ACTION
Change .. 78–C
Delete ...... D

Received by: name _____
title _____
Inspected by: name _____

* Critical Items Requiring Immediate Attention.   Remarks on back (80–1)

FIGURE 5.3 Foodservice inspection report. (*Source:* U.S. Food and Drug Administration)

are weighted from 1 to 5, depending on the importance and severity of the infraction as determined by the FDA. Items marked with an asterisk (*) at the left of the column indicate the items that require immediate action to remedy.

The results of an inspection become public record. In some areas, local newspapers obtain the results and publish them, which evokes mixed reactions from foodservice operators. The results of a satisfactory or exemplary inspection provide "free advertising" for the operation. The operators of establishments that receive less than satisfactory ratings generally complain about the unfairness of the inspections. Such disclosure of inspection results to the local dining public serves as another important reason for foodservice establishments to maintain high standards.

Because a sanitarian visits the average operation only once or twice a year, the results of the inspection indicate the condition of the operation during a very limited period of time, and often may not be an accurate reflection of the cleanliness of the operation. The date and time of the sanitarian's visit in most areas is unannounced, as long as the board of health has not received a complaint which could trigger an immediate inspection to investigate. The components of the inspection report should be included in the regular inspections by the manager so that the operation is always ready for an inspection.

Some areas require that foodservice operators, as well as foodservice employees, receive training and certification in sanitation procedures in order to operate or work in a foodservice operation. The purpose of such a requirement is to raise the awareness of people involved in food handling and service so as to reduce problems and promote food safety. Remember, the board of health visits only once or twice a year, and although the manager is on the premises continuously, he or she cannot be in the kitchen at all times, so if all food handlers receive sanitation training, problems are less likely to arise.

MONETARY REASONS

Maintaining a clean operation and proper sanitary procedures makes sense financially. An operation that covers and dates food stored in the refrigerator and properly rotates foods saves in two ways. First, these procedures will not only help to prevent the growth and spread of bacteria and limit spoilage, they will also reduce waste of valuable food. Second, equipment and machinery that are properly cleaned and maintained will last longer while helping to reduce the spread of harmful bacteria.

There are other monetary reasons to consider. If there is an outbreak of food-borne illness, the operation can lose and, as a consequence, revenue. Along with the short-term reduction in revenue is the probable long-term effect of tarnishing the reputation of the establishment. The two worst advertisements for a foodservice operation are an ambulance parked in front to take a customer or customers to the hospital, and a story in the local news media about a food-borne illness outbreak traced to that place of business. Customers' attitudes are generally negative: "Why take a chance at eating at a place that has had a food-borne illness outbreak?" Such reactions, coupled with the associated legal and medical costs, constitute a major cause of operations going out of business following an unfortunate outbreak of food-borne illness.

An operation to which food-borne illness has been traced is plagued with increased insurance premiums, the potential for expensive law suits, and legal fees. The operation also risks losing its best employees, who are likely to migrate to busier restaurants once the operation is implicated in an outbreak and business begins to suffer. Employees who rely on tips generally cannot afford to stay remain, waiting for the business to rebound.

The loss of customer confidence and goodwill is hard to measure in exact monetary terms, but both are valuable assets to an operation. A customer's choice to dine at a particular place is a vote of confidence in that operation. Such a choice is usually based on a number of factors: price, location, menu variety, and cleanliness. Keep in mind that in the survey results mentioned earlier, dining customers rate cleanliness as one of the most important considerations when choosing where to spend their money.

MORAL REASONS

In addition to the legal and monetary reasons for maintaining sanitary conditions, commercial foodservice operations that serve the public have a moral obligation to provide food that is safe from harmful bacteria and contaminants. People who come to a foodservice operation to spend their money and eat trust that the operators keep the place clean and the food safe.

PROVIDING SAFE FOOD

In order to provide food that is safe to eat, a foodservice operator must be aware of a number of things, including the definitions of *food-borne illness* and *food-borne illness outbreak,* the hazards to food safety and opportunities for contamination, the goals of an effective sanitation program, what bacteria need to survive, and, finally, the eight most commonly cited factors involved in food-borne illness outbreaks.

THE HAZARDS TO FOOD SAFETY

There are three groups of hazards to food safety that are responsible for food-borne illness outbreaks: biological, chemical, and physical (see Figure 5.4). **Biological hazards** are disease-causing bacteria, viruses, and parasites that can be supported by or transmitted in food. **Chemical hazards** involve the contamination of food with toxic substances, such as cleaning and sanitizing chemicals and other substances. **Physical hazards** comprise the contamination of food with nonfood items, such as pieces of glass, staples from cartons, or metal shavings from can openers.

Hazards to Safe Food
> Biological
> Chemical
> Physical

FIGURE 5.4 Hazards to safe food.

THE GOALS OF AN EFFECTIVE SANITATION PROGRAM

The goals of an effective sanitation program, one that ensures that the operation will serve safe food, are relatively simple and straight-forward. They are (1) to protect food from contamination through safe handling procedures and (2) to reduce the effects of contamination.

WHAT BACTERIA NEED TO SURVIVE

If one wants to control the growth and spread of bacteria in order to provide safe food, it is necessary to understand what bacteria require to grow. Different bacteria grow best in different living conditions. The three factors that have the greatest influence on bacterial growth in food are the amount of moisture, availability or lack of oxygen, and the amount of time the food is kept at a temperature that is conducive to such growth.

MOISTURE

Bacteria require an adequate amount of moisture to survive. Because they are microorganisms, they cannot ingest their food—they must receive it in the form of liquid. If the amount of moisture is lowered, bacterial activity is reduced and will eventually stop. This principle makes drying an effective form of preservation for some foods. For example, milk will normally spoil in a short period of time at room temperature, but in powdered milk the moisture is evaporated away and the product will not experience bacterial problems until moisture is replaced. Freezing is also effective in decreasing moisture; bacteria have trouble using it in this form.

PROPER TEMPERATURE/TIME

Most organisms, including bacteria, function most effectively within a certain range of conditions. The bacteria that have the most effect on food grow best in the 60°F to 110°F temperature range. Once in this temperature range, the bacteria must have enough time to multiply to significant numbers in order to cause a problem. Some bacteria can and will grow in a freezer at 0°F, and some will survive and grow in a holding box at 155°F, but most bacterial action occurs within the range called the **temperature danger zone (TDZ)**, 45°F to 140°F. Note that both room temperature and body temperature are within the TDZ, which is why these bacteria cause so many problems in food. Food spends time at room temperature while being prepared, and once it enters the human body the temperature is very conducive to growth (see Figure 5.5).

Foodservice personnel must be especially aware of the time/temperature principle with regard to certain food preparation techniques. One key point is the proper thawing of food—frozen food should thaw under refrigeration or in cold water. Thawing at room

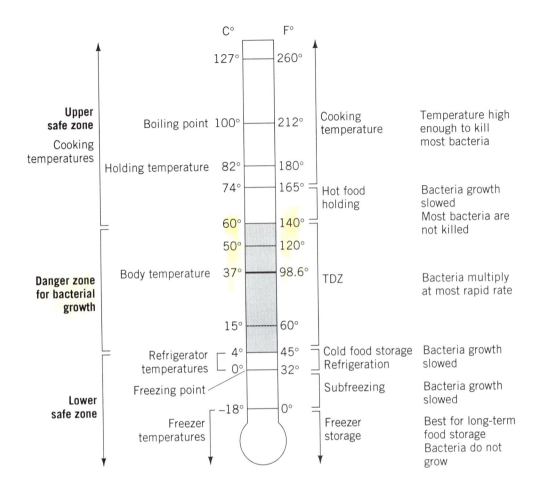

FIGURE 5.5 Critical temperatures.

temperature allows the food to remain in the TDZ too long, thus permitting bacteria to multiply very quickly. Large, dense, frozen products—such as turkeys—that are allowed to thaw at room temperature will thaw on the outside while the inner parts remain frozen.

Different foods support different strains of harmful bacteria, so it is important to cook food items to the proper internal temperature. Poultry should be cooked to 165°F, pork to 150°F, and roast beef to 140°F.

The density of some foods acts as an insulator, reducing the transfer of heat through the food and possibly allowing some foods to remain in the TDZ too long while cooking or cooling. A large pot of chili placed on the stove to heat up exemplifies this problem. The consistency of the chili is so dense that the heat from the stove is not easily transmitted through the food in the cooking vessel. The area on the bottom of the pot near the flame gets very hot, while the chili at the top of the pot remains at room temperature, in the TDZ, allowing bacteria to multiply. A solution to this problem is to stir the product frequently while on it is the stove so that the heat can be evenly distributed, or to heat the chili in several smaller containers to allow better heat distribution.

Holding hot foods for service presents several problems. If a food is kept at a too high temperature, it will continue to cook, causing it to become mushy and overcooked, deteriorate in quality, and possibly dry out. If the food is held at a too low temperature, it becomes a breeding ground for bacteria. A solution is to hold hot food, for a minimum amount of time, at a temperature of 160°F in smaller batches, rather than in one batch for the entire meal period.

AMOUNT OF AVAILABLE OXYGEN

Bacteria are grouped in three categories with regard to oxygen requirements. **Anaerobes** grow only in closed containers where there is no free oxygen. **Aerobes** grow only in environments that have plenty of available oxygen. Most bacteria, however, are in the **facultative** category. These can survive and grow with or without oxygen.

POTENTIALLY HAZARDOUS FOODS

Virtually any food can carry and transmit food-borne illness, although some foods, because of their composition, are more likely to support the growth of harmful bacteria. Given the proper amount of moisture and an appropriate pH level, high-protein foods are classified as potentially hazardous by the U.S. Department of Public Health. The measure of acidity or alkalinity in a food is pH. A food with a pH of below 7 is acidic, while a food with a pH above 7 is alkaline; a food with a pH of 7 is neutral. The ideal range of pH for bacteria growth is 6.6–7.5, while potentially hazardous foods fall into the range of 4.6–7.0. **Potentially hazardous foods** are those that have been traced to the majority of food-borne illness outbreaks. Food products included in this category are those that contain milk or milk products, whole-shell eggs, meats, poultry, fish, shellfish, baked or boiled potatoes, and tofu. These foods require the greatest

Animal Products

Poultry

Fish

Shellfish

Eggs and egg products

Meat

Meat products

Milk and milk products

Prepared Foods

Soups

Custards

Protein salads (with meat or eggs)

Gravies

Potatoes, cooked

Potato salad

Beans or rice, cooked

Sauces

FIGURE 5.6 Some potentially hazardous foods.

attention from foodservice operators. They must be kept refrigerated, owing to their ability to support the rapid growth of disease-causing bacteria (see Figure 5.6).

Compare the previously mentioned foods with some of the other food items used in a kitchen. Think about the food items that are stored at room temperature in foodservice operations—onions, flour, rice, and canned goods. These items can all be stored at room temperature, in the TDZ, as long as they are not altered in any way. This characteristic is due to the nature of the products; some aspect of their composition does not support the growth of bacteria. Flour and rice are safe because of their low amounts of moisture, but after they are mixed with liquid they must be refrigerated. Canned goods are processed in a way that kills all bacteria and then sealed so that the product cannot become reinfected. Once the contents of a can are exposed to air, however, even through only a pinhole, bacteria can be introduced in the product, which must now be kept cool to slow their growth.

THE EIGHT MOST OFTEN CITED FACTORS IN FOOD-BORNE ILLNESS OUTBREAKS

Eight factors are cited most frequently as causes of food-borne illness outbreaks. Awareness of these factors can alert managers to potential problems and allow them to better protect their dining guests.

1. *Failure to cool food properly.* This allows food to remain in the temperature danger zone too long, permitting harmful bacteria to multiply to a harmful level.

2. *Failure to heat or cook food thoroughly.* If food is not brought to a high enough temperature, it remains in the TDZ too long, thus allowing harmful bacteria to multiply.

3. *Employees with illnesses, infections, and/or poor personal hygiene.* Employees who come to work sick or with infected cuts on their hands may transmit harmful bacteria to the food which, in turn, passes the bacteria to the people who eat the food.

4. *Food prepared a day or more before it is to be served.* If food is prepared too far in advance and then held before serving, it will allow bacterial growth. Another problem with preparing food too far in advance is the process of heating-cooling-reheating. Each time food passes through the temperature danger zone it is more susceptible to bacterial contamination.

5. *Contaminated raw ingredients added to ready-to-eat foods.* Foods that are going to be served without heating should not be mixed with raw ingredients. There is a potential problem with this practice: if the raw food is contaminated with harmful bacteria and then added to ready-to-eat food, the bacteria are allowed to go unchecked. The lack of killing heat allows harmful bacteria to multiply quickly.

6. *Foods remaining in the temperature danger zone.* Allowing food, particularly potentially hazardous food, to remain at room temperature or above provides an ideal environment for bacteria to grow and multiply. The elevated temperature in the preparation area increases the risk.

7. *Failure to reheat previously prepared foods.* Previously prepared foods, or leftovers, pose a particular hazard to food safety. Each time a food goes through the TDZ increases the possibility of the accumulation of unsafe levels of bacteria.

8. *Cross-contamination of raw and cooked foods.* Food preparation personnel must be careful not to mix foods. They must also be careful to clean preparation surfaces and tools when switching from one food item to another to minimize the transfer of bacteria, or cross-contamination, from one item to another. For instance, a cutting board and a knife used on chicken that has

been sitting at room temperature must be washed with soap before they can be used to cut vegetables that will be eaten raw.

PROACTIVE APPROACH TO FOOD SAFETY

The foodservice industry has adopted a **proactive** approach to food safety, that is, an approach that seeks to prevent problems before they occur rather than waiting until a problem has occurred and implementing a solution to remedy it. The system promotes self-inspection and is cost-effective for the foodservice operator. The Hazard Analysis Critical Control Points **(HACCP)** system follows the flow of potentially hazardous food through the foodservice area. Using HACCP as a model, the NRA has developed a similar system called Sanitary Assessment of Food Environment **(SAFE)**. The system's goal is to pinpoint potential problem areas and take preventive actions to remedy them.

The HACCP system comprises seven major principles or steps. This system examines and controls the points through which the food goes, from receiving to serving. At any of these points, food can become contaminated and the contaminants can increase and survive (See Figure 5.7).

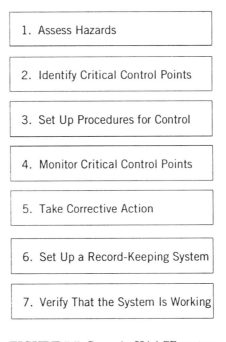

1. Assess Hazards

2. Identify Critical Control Points

3. Set Up Procedures for Control

4. Monitor Critical Control Points

5. Take Corrective Action

6. Set Up a Record-Keeping System

7. Verify That the System Is Working

FIGURE 5.7 Steps in HAACP system.

The Seven Principles of the HACCP System

1. *Assess the hazards at each point in the flow of food through the operation.* Hazards include any biological, chemical, or physical factors that may cause an unacceptable risk to the health of the consumer.

2. *Identify the critical control points.* A **critical control point** is any step in the preparation of an item in which a preventive control can be implemented to eliminate or prevent potential hazards. It is important to observe the preparation, holding, and service processes, as well as employees' care in hand washing and food handling. Flowcharts can be used to map the steps in the processing of food items in order to monitor areas of concern.

3. *Set up procedures for critical control points.* This step involves the establishment of key activities at each control point. These activities should be well defined, observable, and measurable. For the procedures to be effective, employees must be trained properly and provided with the proper tools and equipment.

   The following are examples of control procedures:

   • Washing hands
   • Sanitizing food preparation surfaces and tools
   • Cooking food to specific temperatures
   • Rapid cooling of food

4. *Monitor critical control points.* Using a flowchart (step 2) as a guide, examine each point through which the potentially hazardous food passes, and make sure that the recommendations at each point are being followed.

5. *Take corrective action.* If problems are discovered in the monitoring of control points, quick and corrective action must be taken to remedy the problems. Management must ensure that control procedures are followed; repeated problems or infractions may indicate a need for retraining.

6. *Set up record-keeping procedures.* Documentation should be maintained in order to monitor the success of the system and to provide information to aid in its periodic updating.

7. *Verify that the system is working.* Spot-checks by management must be made to ensure that critical control points are under control. A common misconception is that if a system is in place, it is being used correctly. Management must set up the system, see that it works, and frequently check with production personnel to see that it is being followed.

The HACCP system provides the foodservice operator with a proactive form of self-regulation. Anticipating possible problems gives a manager better control over the mission of providing safe food for guests. The HACCP system also provides material for training new employees in safe food-handling methods.

SUMMARY

Foodservice customers rank cleanliness and sanitation high on their list of priorities when dining out. Managers of foodservice facilities must know how to maintain a clean and sanitary operation. For legal, moral, and monetary reasons, operators must keep the food they serve free from harmful bacteria. To maintain a clean and sanitary operation, foodservice managers must comprehend the key terms used in the field of sanitation, understand the environment that bacteria need to survive, and be aware of the eight most commonly cited causes of an outbreak of food-borne illness so that they can work to avoid them. Armed with this information, foodservice managers can be better prepared to keep an operation clean and free from harmful bacteria and to provide wholesome food for their guests.

KEY TERMS

clean	FDA
contamination	USDA
food-borne illness	wholesomeness
food-borne illness outbreak	CDC
sanitarian	FDA Model Sanitation
sanitary	Ordinance
sanitation	biological hazards
spoilage	chemical hazards

physical hazards
Temperature danger zone
  (TDZ)
anaerobes
aerobes
facultative

potentially hazardous
  foods
proactive
HACCP
SAFE system
critical control point

DISCUSSION AND REVIEW QUESTIONS

1. Explain how an item can be clean but not sanitary.

2. Explain the role of each of the three levels of governmental agencies regarding the safety of the food supply and the cleanliness of foodservice operations.

3. How do local and state health regulations relate to each other? How do state regulations relate to federal standards?

4. Why is it important for a foodservice manager to cooperate with health officials/inspectors when they visit the manager's foodservice operation?

5. How could the FDA Foodservice Establishment Report be used in the daily operations of a foodservice establishment?

6. What are the potential shortcomings of a health inspector's report?

7. What are the three things that bacteria need to survive? Explain what a foodservice operator can do to control each of these three things so as to reduce the growth of bacteria.

8. What makes some foods potentially hazardous and others not?

9. Explain the significance of the HACCP system in a foodservice operation.

10. Why is it important to be proactive rather than reactive when it comes to food safety? Is the HACCP system proactive or reactive? Explain.

11. What is the goal of the HACCP system?

SUGGESTED READINGS

Cichy, R. *Sanitation Management: Strategies for Success* (East Lansing, MI: Educational Institute of the American Hotel & Motel Association, 1993).

Educational Foundation of the National Restaurant Association. *Applied Foodservice Sanitation,* 4th ed. (Chicago: Educational Foundation, 1991).

Loken, Joan K. *The HACCP Food Safety Manual* (New York: John Wiley & Sons, 1995).

CHAPTER SIX

MENU PLANNING

CHAPTER OBJECTIVES

Upon completing the chapter you should be able to:

- Discuss how customers view menus.
- State the role and purpose of the menu in a foodservice operation.
- Discuss how a menu is used as a marketing tool.
- Distinguish between a static and changing menu.
- Describe the characteristics of the five basic types of menus.
- Tell how to use a menu as a selling tool.
- List the common mistakes found on menus.
- List and explain the points that must be considered when planning a menu.

The success of most restaurants is often directly associated with the planning of the menu. When used in conjunction with good service, quality food, and a clean operation, a properly planned menu can help to accomplish the three goals of a successful restaurant:

1. To increase the amount of money the customer spends.
2. To attract new customers and business.
3. To increase the frequency of visits by the customers.

A menu should do more than simply convey what the operation offers for sale. When properly designed, a menu both informs customers and influences their purchasing decisions by serving as a merchandising tool to entice the customer.

The objective of this chapter is to provide an understanding of the purpose and role of the menu in a foodservice operation. Different types of menus for different types of applications are presented, along with the components and uses of each type.

Although a foodservice operation must be customer-driven, the day-to-day workings of the foodservice restaurant are "menu-driven." In planning an operation, the menu is one of the first things to be addressed. Once the market research is completed and the theme of the operation determined, the menu is developed. The menu then dictates the number and types of employees, the size and layout of both the kitchen and the dining room, the style of service, and the equipment.

MENU PLANNING FROM THE CUSTOMER'S POINT OF VIEW

The typical foodservice patron views a menu as more than a mere listing of what an establishment has to offer for sale. The appearance of the menu reveals the professionalism of the operation. A handwritten menu reflects a "homey" atmosphere, whereas one that is professionally typeset indicates a more sophisticated and sometimes elegant atmosphere. A tattered, greasy menu with prices "whited-out" and items crossed out reflects a lack of professionalism. A menu with small print and difficult-to-read writing may cause the customer to feel that the operation is trying to hide something. Menu items hidden under "clip-on" specials cause further problems and confusion for guests.

THE PURPOSE OF THE MENU

A menu serves more than one purpose: it is a source of information and a marketing tool.

THE MENU AS A SOURCE OF INFORMATION

By listing the operation's food and beverage offerings, a menu informs the customer as to what is for sale. The description of each menu item should explain the ingredients and components of the dish.

In addition, a menu sets the tone for the dining experience. It conveys the atmosphere of the establishment—the **theme** and **concept**—along with the items that are offered for sale. An informal menu with bright colors, humorously written and illustrated, conveys a feeling of casual dining. A formal menu, featuring **classical dishes** and generally without illustrations, sets an elegant tone for the establishment and for the dining experience.

THE MENU AS A MARKETING TOOL

A menu must be designed to satisfy the customer. If the restaurant's food and beverage selections satisfy customers, they are likely to return, and the operation will be successful. If the restaurant's offerings do not satisfy customers, they will take their business and their money elsewhere.

The goal of **market research** is to determine what an operation must offer to satisfy its customer base. Customers have diverse tastes in food; one restaurant cannot expect to satisfy all. A target market, or group of consumers, must be determined to help management decide what offerings to provide. Research is done on the demographics (regarding age and economic status) of the local area, and the menus of competitor restaurants are examined.

Once the customer segment is determined, the menu should be planned to target or focus on the market the operation plans to attract. The type and pricing of the menu offerings must be designed to satisfy the target market, along with the hours of operation and the atmosphere of the facility.

Menus can be an effective **marketing tool**. Posting them outside the restaurant can allow potential customers to examine its offerings before entering. An attractive menu design can entice would-be customers to enter. Menus can also be distributed through the local chamber of commerce, tourist information centers, hotels, and motels to help market the restaurant and to draw in customers.

Some dining patrons do not ask questions about items on a menu; if they do not recognize a menu item, in all likelihood they will not order it. These customers may be missing some of the best menu selections because they are too timid to ask questions. Descriptive copy provided below menu items, however, serves to inform and entice the customer.

Descriptive copy is the written description of a menu item, including the ingredients and method of preparation. It does not have to be given for all items on a menu (a mushroom omelet, for example, is self-explanatory).

Some items may be given various designations, such as Chef's Special, House Special, or signature items. Generally, a **signature item** is named after an organization, such as "Marriott Burger," or after the restaurant or hotel, and such names should be reserved for items of high quality that are produced on the premises. Customers may judge the reputation of an operation by these items. In addition, it is important that terms used in merchandising a menu are completely accurate.

THE ROLE OF THE MENU IN THE DAY-TO-DAY OPERATIONS OF THE FOODSERVICE OPERATION

The menu offerings of a foodservice operation dictate the types of foods and beverages that are served. They also determine the kind of equipment required, the necessary skill levels and size of staff—both front- and back-of-the-house—the dining room layout, and the type of service provided (see Figure 6.1).

PRODUCTS ON THE MENU

The variety of products on a menu affects the kinds of foods and beverages served. A large number of items on a menu with varied types of foods may encourage the use of more processed products. For example, french fries are very seldom made from scratch, owing to the convenience and quality of the frozen product. Bakery products may be prepared in-house or purchased ready to serve, depending on management's assessment of customer preferences.

The menu serves as the starting point for the purchasing agent. The operation must ensure that all items listed on the menu are available for sale and that all necessary ingredients are available from its suppliers and vendors.

EQUIPMENT AND KITCHEN SPACE REQUIREMENTS

An extensive menu and greater variation in menu items may require a larger variety of equipment and necessitate more kitchen space. For example, fast-food restaurants have limited menus; therefore, they require a modest amount of equipment and kitchen space. Fine-dining operations with full-service menus have a greater number of menu selections and thus require more kitchen space and more equipment for the various types of pro-

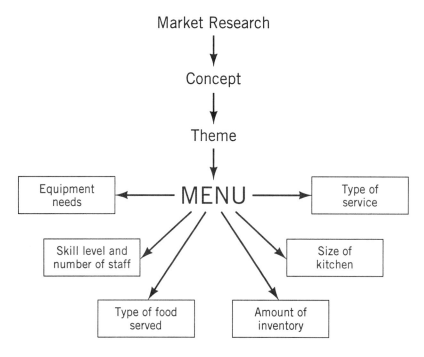

FIGURE 6.1 The role of the menu in a foodservice operation.

duction. Institutions—educational, health care, and military—also require larger kitchen areas and more varied types of equipment to produce a great many types of food.

NUMBER AND SKILL LEVEL OF STAFF

The more complex the menu, the higher the level of staff skill and the greater the number of staff members will be required. Both front- and back-of-the-house personnel are affected. Complex menu items, such as bouillabaisse (a French fish stew), which utilizes a recipe far more difficult than those used for most standard items on a menu, require a trained and high-quality staff. Conversely, a menu item such as a hamburger is rather easy to prepare and requires a staff with a minimal amount of training.

STATIC AND CHANGING MENUS

Menus may be either static or changing. A **static menu** offers the same items daily. A drawback of this type of menu is that because the menu does not change, the purchaser is generally unable to take advantage of special deals offered by suppliers and seasonal specialties. Moreover, if a static menu item is not selling, the restaurant is stuck with the ingredients in inventory until the menu is changed.

An advantage of a static menu is that a customer's choice of a place to dine is made easier, because the customer knows what menu selections to expect. Static menus work well in restaurants and other operations that cater to a broad-based clientele. However, when a menu remains constant over time, regular diners may grow bored with the selections offered.

A **changing menu** changes its offerings from time to time, sometimes at regular intervals. Changing menus usually cater to a regular clientele. Operations that serve the same clientele daily (for example, employee cafeterias, educational foodservices, and long-term medical facilities) should occasionally alter their menu offerings to satisfy their customers.

TYPES OF MENUS

There are five basic types of menus: à la carte, du jour, cyclical, table d'hôte, and limited.

A LA CARTE

An **à la carte** menu is a static menu in which selections are priced separately, whereby the customer selects his or her items individually. A la carte menus are popular in fast-food restaurants, snack bars, and cafeterias. They are also found in table-service restaurants but are generally decreasing in popularity in this segment of the foodservice industry. Some customers prefer the opportunity to choose their own combination of items, although the general perception of an à la carte menu in a table-service restaurant is that the total meal will be more expensive than a combination dinner created by the restaurant. Generally, most table-service restaurants' menus have à la carte sections from which guests can supplement their meal selections.

DU JOUR

Du jour menus feature items that change on a daily basis. The operation has an opportunity to take advantage of food products in their peak season when quality is at its best. The opportunity to change the menu daily gives the purchaser the ability to select "specials" from suppliers.

An operation that features this type of menu requires a more creative and experienced chef. The guests who patronize an operation with a *du jour* menu are typically drawn there because of the reputation of its chef. Generally, customers are not aware of what the day's offerings are until they arrive at the establishment.

Du jour menus may be posted on a chalkboard or printed daily by word processor and distributed to the guests. The operation is not locked into any regular menu items, so if the price of any ingredients rise or if an item does not sell well, the item may simply be excluded from future menus (see Figure 6.2).

CYCLICAL

A **cyclical menu** is a type of changing menu that rotates selections over a period of time. Such menus are primarily used in institutional operations and work best in those that serve a captive audience such as in a school, where there is need for the menu to change daily to avoid monotony. A cyclical menu generally repeats every three, six, or eight weeks, but the more popular items may be repeated several times within the cycle. For example, hamburgers and pizza may be repeated two or three times within a college foodservice operation cycle, whereas pot roast may only appear once in the same cycle.

Proper production planning for a cyclical menu should keep the amount of leftovers to a minimum. Because the same customers dine in the typical institutional operation daily, leftovers cannot be served a second time without some changes. The sequence of the meals in a cycle menu should be planned so as to use any leftovers. For example, the cooked hamburger patties left from dinner one night may be ground up for the next night's tacos; the leftover baked chicken used as an entree for one meal can be skinned, boned, and diced to be used in the following day's chicken soup.

A drawback of a cyclical menu for a foodservice operation is that the diversity of the menu selections offered in the cycle require a wide variety of equipment, a large preparation area, and sizable storage areas. There are also advantages to the cycle menu. Through the repetition of the cycle and the duplication of preparing the menu items, the preparation staff becomes familiar with the recipes, and statistics can be derived as to the popularity of various items so that

SANDWICHES

Certified Chuck Burger *with Monterey Jack Cheese*	6.95
Add Peppers and Onions	7.50
Grilled Chicken Breast *with Roasted Sweet Peppers,*	
Fontina Cheese and Basil Aioli	6.95
Mediterranian Pita *with Hummus, Tomato - Cucumber Salad,*	
Feta Cheese and Olives	6.50
House Smoked Turkey *on Marble Rye*	
with Swiss Emmenthaler and Beefsteak Tomatoes	6.75
Spit Roasted Lamb *on Rosemary Foccacia*	
with Roasted Garlic and Brie Sauce	7.95

PASTAS

Cheese Tortellini *with Proscuitto, Sundried Tomatoes and Cream*	10.95
Rigatoni *with Spit Roasted Chicken, Wild Mushrooms, Marsala and Sage*	10.95
Linguini *with Duck Sausage, Artichoke Hearts,*	
Prosciutto and Jalapeno - Garlic Cream Sauce	10.95
Black Pepper Fettuccine *with Andouille Sausage, Smoked Duck and Shrimp*	11.50
Saffron Risotto *with Prawns, Scallops and Tomatoes*	11.95

SEAFOOD

Grilled Bacon Wrapped Thresher Shark Brochette	
with Tomatoes, Capers, Basil and Virgin Olive Oil	10.75
Grilled Blackened Catfish Filet	
with Herb - Mustard Sauce and Two Salsas	9.95
Bamboo Steamed King Salmon and Shrimp	
with Mixed Vegetables, Pasta Noodles and Yogurt Dill Sauce	11.95
Grilled California Rock Cod	
with Tomato Creme Fraiche and Chives	9.75

FIGURE 6.2 A *du jour* menu.

the operation can better forecast the menu mix and plan production for the future (see Figure 6.3).

TABLE D'HÔTE

A **table d'hôte menu** includes all the specific courses of the meal—the appetizer, the entree, a vegetable, a starch, and dessert—at a fixed (set) price. Menu selections are decided by the chef, who takes into account both balance of color and flavor combinations. The menu is set, and customers generally cannot make substitutions for items on the menu.

CYCLE 1   MENU
GUEST PRICES: BREAKFAST $1.75   LUNCH $4.25   DINNER $5.50

UNIVERSITY DINING SERVICES

MONDAY 10/16	TUESDAY 10/17	WEDNESDAY 10/18	THURSDAY 10/19	FRIDAY 10/20	SATURDAY 10/21
Breakfast Juice	Breakfast Juice	Breakfast Juice	Breakfast Juice	Breakfast Juice	BUTTERFIELD HALL
Fruit/Cereal/Yogurt/	Fruit/Cereal/Yogurt/	Fruit/Cereal/Yogurt/	Fruit/Cereal/Yogurt/	Fruit/Cereal/Yogurt/	CONTINENTAL BREAKFAST
Bread Bar	Bread Bar	Bread Bar	Bread Bar	Bread Bar	8:30 A.M. – 10:00 A.M.
Fried Eggs	Hard-Cooked Eggs	Scrambled Eggs	Fried Egg/Biscuit	Poached Eggs	Assorted Juice and Fresh Fruit
Blueberry Pancakes	French Toast/Syrup	Pancakes/Toppings	Toast-Own-Waffles	French Toast/Syrup	Cinnamon Rolls
Home-Fried Potatoes	Crisp Bacon Slices	Cottage Fry Potatoes	Canadian Bacon	Corned Beef Hash	Bagels/Cream Cheese
Fresh Baked Muffin	Cranberry Nut Bread	Fancy Doughnuts	Strawberry Coffee Cake	Blueberry Bread	English Muffins
Bagels/English Muffins	Bagels/English Muffins	Bagels/English Muffins	Bagels/English Muffins	Bagels/English Muffins	Toast/Butter/Jelly
Toast/Butter/Jelly	Toast/Butter/Jelly	Toast/Butter/Jelly	Toast/Butter/Jelly	Toast/Butter/Jelly	Beverages
Beverages	Beverages	Beverages	Beverages	Beverages	
Broccoli Cheese Soup	Country Vegetable Soup	Poor Boy Soup	Mulligatawny Soup	Manhattan Clam Chowder	LUNCH
Chicken Pattie/Roll	Grilled Cheese or	Pizzarino Sandwich	Shaved Steak/Cheese	Open-Faced BLT	11:00 A.M. – 2:00 P.M.
American Chop Suey	Grilled Ham & Cheese	Turkey Pot Pie	Grinder	Baked Macaroni & Cheese	Oriental Noodle Soup
Chopped Spinach	Chinese Pepper Steak	Spinach Egg Roll	Chicken Croquettes/	Chocolate Pudding	Made-to-Order
Ice Cream Novelties	Whole Kernel Corn	Devil's Food Cake	Gravy		N.Y.-Style Deli
	Strawberry Shortcake		Chuckwagon Blend Veg.		Sandwich
			Ranger Cookie		Frankfurt/Roll
					Breaded Mushrooms
					Assorted Desserts
FAST FOOD SELECTIONS	FAST FOOD SELECTIONS	FAST FOOD SELECTIONS	FAST FOOD SELECTIONS	FAST FOOD SELECTIONS	
Hope Dining Hall	Hope Dining Hall	Hope Dining Hall	Hope Dining Hall	Hope Dining Hall	
DELI SELECTIONS	DELI SELECTIONS	DELI SELECTIONS	DELI SELECTIONS	DELI SELECTIONS	
Butterfield/Roger Wms	Butterfield/Roger Wms	Butterfield/Roger Wms	Butterfield/Roger Wms	Butterfield/Roger Wms	
Soup de Jour	Soup de Jour	Soup de Jour	Soup de Jour	Soup de Jour	DINNER
Shrimp Tahiti/Tartar	Meatloaf w/Gravy	Fresh Filet of Haddock	Stir-Fry Beef w/	Lasagna	4:30 P.M. – 7:00 P.M.
Sauce	Honey-Glazed or Baked	w/Lemon Sauce	Broccoli	Fish Sticks/Tartar	CHINESE SPECIALTIES
Tortellini w/Sauce	Chicken Quarter	BBQ Spare Ribs	Scallops Broiled in	Sauce	
Skin-on Potato Wedges	Vegetarian Cheese/	Cheese Quiche	Garlic Butter	Vegetarian Lasagna	
Vegetarian Cheese/	Spinach Casserole	Curried Rice	Vegetarian Beans/Rice	Mashed Potatoes	
Spinach Casserole	O'Brian Potatoes	Self-Serve Vegetable	Egg Noodles	Self-Serve Vegetable	
Self-serve Vegetable Bar	Self-Serve Vegetable	Bar	Self-Serve Vegetable	Bar	
Fresh Baked Rolls	Bar	Fresh Baked Rolls	Bar	Fresh Baked Rolls	
Peanut Butter Cake	Fresh Baked Rolls	Ice Cream Smorgasbord	Fresh Baked Rolls	Apple Pie	
	Butterscotch Brownie		Marble Cake		
HOPE DINING HALL	HOPE DINING HALL	HOPE DINING HALL	HOPE DINING HALL	HOPE DINING HALL	
Chicken/basket	Italian Specialties	Hot Roasted Sandwiches	Italian Specialties	Closed	
BUTTERFIELD HALL	BUTTERFIELD HALL	BUTTERFIELD HALL	BUTTERFIELD HALL	BUTTERFIELD HALL	
Deli Express	Deli Express	Deli Express	Deli Express	Deli Express	

SALAD BAR, JELLO, FRESH FRUIT, PEANUT BUTTER, JELLY, AND BEVERAGES ARE AVAILABLE AT ALL LUNCHEON AND DINNER MEALS

MENU SUBJECT TO CHANGE

FIGURE 6.3 A cyclical menu.

CYCLE 2   MENU   UNIVERSITY   DINING SERVICES

GUEST PRICES: BREAKFAST $1.75   LUNCH $4.25   DINNER $5.50

MONDAY 10/23	TUESDAY 10/24	WEDNESDAY 10/25	THURSDAY 10/26	FRIDAY 10/27	SATURDAY 10/28
Breakfast Juice Fruit/Cereal/Yogurt/ Bread Bar Scrambled Eggs Toast-Own-Waffles Baked Ham Slice Fresh Baked Muffin Beverages	Breakfast Juice Fruit/Cereal/Yogurt/ Bread Bar Fried Eggs French Toast Cottage Fry Potatoes Fancy Doughnuts Beverages	Breakfast Juice Fruit/Cereal/Yogurt/ Bread Bar Hard-Cooked Eggs Pancakes/Syrup Pork Sausage Links Streusel Coffee Cake Beverages	Breakfast Juice Fruit/Cereal/Yogurt/ Bread Bar Scrambled Eggs Toast-Own-Waffles Crisp Bacon Slices Pumpkin Bread Beverages	Breakfast Juice Fruit/Cereal/Yogurt/ Bread Bar Fried Eggs Blueberry Pancakes/ Syrup Hash Brown Potato Pattie Fresh Baked Muffin Beverages	BUTTERFIELD HALL CONTINENTAL BREAKFAST 8:30 A.M. – 10:00 A.M. Assorted Juice and Fresh Fruit Cinnamon Rolls Bagels/Cream Cheese English Muffins Toast/Butter/Jelly Beverages
Minestrone Soup Open-face Hot Beef Sandwich/Gravy Chicken Fingers/ Dipping Sauce Seasoned Peas Fruit Harmits	Chicken Rice Soup Grilled Pastrami on Kaiser Roll Cheese/Pepperoni Boboli Pizza Breaded Zucchini Sticks Congo Bars	Beef Noodle Soup Italian Sausage on Grander Roll Vegetable/Cheese Quiche Country Blend Veg. Sherbet Cups	Chicken Escarole Soup Crumb-Topped Light Filet of Cod Mexican Tacos Broccoli Spears Roman Apple Cake	N.E. Clam Chowder Grilled Cheese Sandwich Beef Steak/Biscuit Green Beans Piquant Vanilla Pudding	LUNCH 11:00 A.M. – 2:00 P.M. Cream of Spinach Soup Made-to-Order N.Y. Style Deli Sandwich Meatball Grinder Breaded Mozzarella Sticks Assorted Desserts
FAST FOOD SELECTIONS Hope Dining Hall DELI SELECTIONS Butterfield/Roger Wms	FAST FOOD SELECTIONS Hope Dining Hall DELI SELECTIONS Butterfield/Roger Wms	FAST FOOD SELECTIONS Hope Dining Hall DELI SELECTIONS Butterfield/Roger Wms	FAST FOOD SELECTIONS Hope Dining Hall DELI SELECTIONS Butterfield/Roger Wms	FAST FOOD SELECTIONS Hope Dining Hall DELI SELECTIONS Butterfield/Roger Wms	
Soup de Jour Veal Parmigiana Lemon Rice Stuffed Filet of Flounder Fettucine Alfredo Shell Macaroni Self-Serve Vegetable Bar Fresh Baked Rolls Ice Cream Novelties	Soup de Jour Roast Turkey/Gravy Dressing/Cranberry Sauce Fish Munchies Tofu Stuffed Noodles Mashed Potatoes Self-Serve Vegetable Bar Fresh Baked Rolls Blueberry Pie	Soup de Jour Shepherd's Pie Chinese Chicken Wings Zucchini/Feta Cheese Vegetarian Casserole Pork Fried Rice Self-Serve Vegetable Bar Fresh Baked Rolls Cocoanut Cream Pie	Soup de Jour Roast Fresh Ham/Gravy Applesauce Meatloaf/Gravy Brown Rice/Pea Vegetarian Casserole Oven Brown Potatoes Self-Serve Vegetable Bar Fresh Baked Rolls Chocolate Chip Cookie	Soup de Jour Lemon Baked Chicken Quarter/Gravy Cranberry Sauce Seafood Stuffed Filet of Sole/Cheese Sauce Vegetarian Chili Baked Potato/Sour Cream Self-Serve Vegetable Bar Fresh Baked Rolls Spice Cake/Frosting	DINNER 4:30 P.M. – 7:00 P.M. RECIPES FROM HOME
HOPE DINING HALL Chicken/basket BUTTERFIELD HALL Deli Express	HOPE DINING HALL Italian Specialties BUTTERFIELD HALL Deli Express	HOPE DINING HALL Hot Roasted Sandwiches BUTTERFIELD HALL Deli Express	HOPE DINING HALL Italian Specialties BUTTERFIELD HALL Deli Express	HOPE DINING HALL Closed BUTTERFIELD HALL Deli Express	

MENU SUBJECT TO CHANGE

SALAD BAR, JELLO, FRESH FRUIT, PEANUT BUTTER, JELLY, AND BEVERAGES ARE AVAILABLE AT ALL LUNCHEON AND DINNER MEALS

FIGURE 6.3 (continued)

A table d'hôte menu generally includes five to seven courses. Skill and experience are needed to plan the multiple-course menu so that the foods complement each other. The menu planner must be very attentive to the food preferences of the clientele so that a fixed menu that will sell can be developed (see Figure 6.4).

LIMITED MENU

A **limited menu** is static and generally includes only six to eight main course items and a limited offering of accompaniments. This type of menu is successful in both fast-food restaurants and specialty restaurants such as steak and seafood houses.

There are advantages for the foodservice operation in reducing the number of menu items offered for sale. The amount of food required in inventory is reduced, which in turn reduces the amount of money tied up in inventory. The usage of individual food items is thus increased and, therefore, the chances of food spoilage and pilferage are also minimized.

Another advantage of having few items on a menu is that it is easier for both the preparation and the serving staff to become more familiar with the menu items, thereby possibly reducing the number of errors by back-of-the-house personnel. Servers likewise have fewer menu items to memorize, so they can better explain the features of the items to customers. A limited menu also affects other aspects of the foodservice operation: less equipment is required, storage space is reduced, and smaller back-of-the-house preparation areas are needed (see Figure 6.5).

THE COMBINATION OF MENU TYPES

Most menus in table-service restaurants contain both portions that are static and portions that are changing, along with components of the five variations of menus. For example, the main portion of a **combination menu** may be static, with a modified table d'hôte section in which entrees are served with other courses, such as a salad, and various accompaniments such as vegetables and starches. "Daily specials" are generally offered on the menu, which rotate without any set schedule, as in a *du jour* menu, and other specials on the menu may rotate over a fixed schedule, as in a cyclical menu. The menu may also include combination meals for a fixed price, as in a table d'hôte menu, and

Le Dîner

Hors d'Oeuvres
Choice of One

Sauteed Oysters
laced with a ginger cream sauce

Black Angus Beef Carpaccio
enriched with extra virgin olive oil and fresh cracked pepper

Cassolette of Escargots
adorned with shitaké mushrooms and coriander butter

Soupe du Jour
Froide ou Chaude
New England Clam Chowder

Fruit Gazpacho

La Salade

An Array of Mixed Baby Greens
tossed with roasted baby corn and southwestern salsa

Le Granite

Lemon Lime Basil

Les Entrées
Choice of One

Braid of Petrale Sole and Silver Sterling Salmon
braised in champagne cream sauce

Pan Seared Moulard Duck Breast
mantled with mango aigre-doux and fresh fanned mango

Grilled Alaskan Halibut
glazed with a Méditérranéan hollandaise and saffron rice

Sauteed Noisettes of Lamb
garnished with a pear croquette and tarragon jus

Seared Jumbo Sea Scallops Set on a bed of Fresh Fennel Julienne
atop tomato angel hair pasta mantled with a dill vinaigrette

Lightly Breaded Veal Oscar
crowned with crab meat, asparagus and bearnaise

A La Facon Du Chef
a main course prepared spontaneously by the chef

Forty-Five Dollars
Lundi le 7 Octobre

Gratuity Not Included

Les Patisseries

Assortment of Pastries & Desserts
Evian Natural Spring Water · $3.75

FIGURE 6.4 A table d'hôte menu.

Available Lunch and Dinner

HAMBURGERS

Fresh chuck, mesquite grilled to your preference, served on a toasted sesame seed bun or onion roll. Includes your choice of french fries, coleslaw, red potato salad or fresh fruit.

THE BURGER 4.95

CHEESEBURGER 5.25
American, swiss or cheddar

**MUSHROOM—
SWISS CHEESEBURGER** 5.75

**GUACAMOLE
CHEESEBURGER** 5.75

PATTYMELT 5.50
Grilled on rye, with grilled onions

**BACON CHEDDAR
BURGER** 5.75

**BEARNAISE SAUCE
BURGER** 5.95
With grilled onions and mushrooms

BELTBUSTER 8.50
An enormous one-pound gourmet special. We take your photograph to commemorate the occasion.

SANDWICHES

Served with your choice of fresh fruit, coleslaw, red potato salad or french fries.

PRIME RIB DIP 6.95
Thinly sliced prime rib, mushrooms and melted swiss cheese, served with au jus

TRIPLE DECKER 5.75
Sliced turkey breast, canadian bacon, avocado, lettuce and tomato on rye

FILET MIGNON 8.50
Served on toasted sour dough

**SMOKED
TURKEY BREAST** 6.25
Melted havarti cheese, tomatoes and pesto mayonnaise on toasted sour dough

**SOUTHWESTERN
PRIME RIB DIP** 6.95
Thinly sliced prime rib, green chiles and melted cheddar cheese, served with au jus

SALADS

FAJITA SALAD 6.50
Blue and white tortilla chips, mesquite grilled steak or chicken, cheddar cheese, tomatoes, guacamole, olives, sour cream and salsa

CHICKEN SALAD 5.75
Fresh chicken tossed with seedless grapes and toasted pecans, served with fresh fruit and a croissant

BUSTER SALAD 5.75
Mixed greens, smoked ham, genoa salami, swiss cheese, provolone, avocado, turkey breast, cucumbers and tomatoes

**TOSSED GREEN, FRESH
SPINACH OR CAESAR
DINNER SALAD** 2.75

FIGURE 6.5 A limited menu.

appetizers, accompaniments, and desserts can be listed and priced separately in the fashion of an à la carte menu (see Figure 6.6).

USING THE MENU AS A SELLING TOOL

The average foodservice consumer spends only about three minutes examining a menu. With that in mind, the menu planner can use some techniques to highlight menu items so as to influence what the customer orders.

The menu as a selling tool can influence the customer's buying decisions in three ways:

1. By offering suggestions
2. By highlighting specials items
3. By the arrangement of the menu items on the page

OFFER SUGGESTIONS ON THE MENU

A menu can be an excellent selling tool to supplement the suggestive selling of the front-of-the-house staff. One reason that menus are so effective is that some customers are leery of asking questions or asking for menu suggestions. The foodservice operation can increase sales and customer satisfaction by including combination dinners on the menu and incorporating suggestions for their dining patrons.

A menu can include recommendations for wines or other items that complement the menu selections. Accompaniments—such as a scoop of ice cream with a piece of pie or a slice of cheese on a plain hamburger—will increase guests' dining satisfaction and increase the operation's sales revenue.

HIGHLIGHT SPECIAL ITEMS

Because customers spend such a short time scanning a menu, planners use several techniques to emphasize the items they want to catch the customer's eye. Circling a selection on the menu or placing an illustration next to it can draw the customer's attention. Adding the word "Special" in the margin beside a selection is also an effective means of highlighting a menu item.

WEEKDAY BREAKFAST SPECIALS

Choose any breakfast below Monday-Friday til 11 a.m. and pay just 3.49

• Cheese Omelette   • The Big Two   • The Traditional   • Strawberry Pancakes

FROM THE GRIDDLE

Strawberry French Toast   4.70
Thick, crispy french toast topped with strawberries and whipped
cream and served with ham, bacon or sausage.
    Plain French Toast with ham, bacon
    or sausage served with warm syrup.   4.15
    Plain French Toast with warm syrup.   3.15

Blueberry Pancakes   4.50
Three fluffy pancakes with blueberries inside. Served with ham,
bacon or sausage.
    Blueberry Pancakes with warm syrup.   3.50

Strawberry Pancakes   4.50
Three fluffy pancakes topped with strawberries and whipped
cream. Served with ham, bacon or sausage.
    Strawberry Pancakes   3.50
    Plain Pancakes with ham, bacon
    or sausage served with warm syrup.   3.95
    Plain Pancakes with warm syrup.   2.95

BREAKFAST FAVORITES

 # 1 Just One   3.55
One egg, any style; your choice of ham, bacon or sausage;
breakfast potatoes and toast.

 # 2 The Traditional   4.25
Two eggs, any style; your choice of ham, bacon or sausage; served
with breakfast potatoes and toast.

 # 3 Pancake and Egg Combo   4.55
Two eggs, any style; three buttermilk pancakes and your choice
of ham, bacon or sausage.

 # 4 French Toast Combo   4.79
Two eggs, any style; four pieces of our thick french toast with
choice of ham, bacon or sausage.

 # 5 Steak and Eggs   6.95
A delicious steak, cooked to your order; two eggs, any style;
served with breakfast potatoes and toast.

 # 6 Country Fried Steak 'N Eggs   5.50
Country fried steak topped with home style gravy, two eggs,
served with breakfast potatoes and toast.

 # 7 Ham Quickie   3.95
Two eggs scrambled with diced ham and served with breakfast
potatoes and toast.

 # 8 The Big Two   3.90
Two eggs, any style, two strips of bacon, two sausage links and
two pancakes.

*English muffin or buttermilk biscuit may be substituted for
toast or choose a bagel instead for .30 additional.*

OMELETTES

Western   4.75
A classic — filled with a blend of diced ham, bell peppers and
onions. Served with seasoned breakfast potatoes and your choice
of English muffin, white, wheat or rye toast.

Supreme   4.95
Filled with real bacon bits, sauteed onion and melted Swiss
cheese. Served with seasoned breakfast potatoes and your choice
of English muffin, white, wheat or rye toast.

Ham and Cheese   4.65
Filled with savory diced ham and melted American cheese.
Served with seasoned breakfast potatoes and your choice of
English muffin, white, wheat or rye toast.

Cheese   4.35
A cheese lover's delight — filled with American or Swiss cheese.
Served with seasoned breakfast potatoes and your choice of
English muffin, white, wheat or rye toast.

SIDE ORDERS

Ham, Bacon or Sausage	1.85
Breakfast Potatoes	1.15
Toast, English Muffin or Biscuits	.95
Freshly Baked Muffin	.75
Bagel with Cream Cheese	1.50
Hot Cinnamon Roll	1.25
Assorted Cold Cereals with fruit	1.15
Quaker® Oatmeal with toppings	1.15

BEVERAGES

	Large	Regular
Orange Juice	1.35	.99
Apple, Tomato or Grapefruit Juice	1.35	.99
Milk, Chocolate Milk		.99
Hot Chocolate		.89

*We feature unlimited refills for
freshly brewed coffee, decaffeinated coffee
and hot tea.*

Low cholesterol **egg beaters,**
are available upon request.

We serve only 100% cholesterol-free margarine.

Please ask your server for our children's menu.

FIGURE 6.6 A combination menu.

PLAN THE ORDER OF ITEMS ON THE MENU

Most menus separate their offerings into lists of similar items, such as appetizers, salads, entrées, and desserts. The placement of items in the list affects the frequency with which they are ordered. The menu items located in the first and last positions on the list are generally ordered most frequently. These two premium positions should be reserved for menu items that are the easiest for the operation to prepare and that provide the highest profit. Items positioned in the middle of a list sell with the least frequency. Those that are harder to prepare or that do not generate a high profit should be placed in these positions.

DESIGNING THE MENU

TEAM APPROACH IN MENU DESIGN AND EVALUATION

To increase a menu's effectiveness and to broaden its appeal, it should be designed and evaluated by a team of people in the foodservice operation. A team approach helps to eliminate bias. The team should include personnel who represent both front-of-the-house and back-of-the-house activities.

Food servers, who spend more time with customers than other staff members, are most aware of the customers' preferences. Servers are an integral component of the menu design and evaluation team. The chef/kitchen manager and purchaser must also be included on the team. The chef provides input on the feasibility of producing the menu items. The purchasing agent provides important information on ingredient availability, cost, and seasonal items. The team should meet periodically to evaluate the present menu's effectiveness and the customers' comments on the current selections.

HOW OFTEN THE MENU SHOULD BE CHANGED

Seasonal eating patterns often dictate menu changes. Hearty soups and heavy entrées sell well during cold winter months. Conversely, customers look for menu items that are lighter and more refreshing in hot summer months.

Printing and producing a multiple-color or pictorial menu is expensive. Menu items selling at prices that do not reflect recent rises in ingredient costs result in lost revenue for the operation. The cost of reprinting the menu must be balanced against the revenues lost by an outdated menu.

The menu should be evaluated periodically to see how it compares with the competition and how menu items are selling. Ingredient prices are volatile and hard to predict. When ingredient prices increase or product sales begin to decline, it is time to consider revising the menu.

Management must keep an open line of communication with customers. The use of suggestion boxes and guest comment cards and talking to customers are effective ways to monitor customers' feelings about an operation and its menu offerings. When regular patrons of a foodservice operation begin to get bored with its offerings, the time has come for management to reevaluate the menu. If an operation cannot keep its customers interested in its menu selections, the customers may go elsewhere to dine.

COMMON MISTAKES IN MENUS

There are items that appear on menus that reduce the menus' ability to sale and merchandise the items of the operation. The commonly accepted "mistakes" and their explanations should be examined by future managers to judge their affect on the "saleability" of the menu.

LACK OF SPECIALS

Regular patrons of a foodservice establishment generally look for "specials of the day" to reveal the quality of the kitchen personnel and to provide some variation from the regular menu. Specials allow the kitchen personnel to exhibit their creativity, and if a special sells well, it may earn a permanent place on the menu.

MENUS THAT ARE TOO CROWDED

When menu items are placed too close together, when clip-on notes about Specials cover other items, or when the print or handwriting on the menu is hard to read or understand, customers may be confused and may not order as much as they would if the menu were easy to read. Special care should be taken to ensure that all offerings on a menu are easy to read and that items on the menu do not cover one another. Desserts and beverages should not be hidden on a back page, obstructed from the view of all but the most curious customer.

USE OF UNCOMMON TERMINOLOGY WITHOUT CLARIFICATION

Culinary terminology may be unfamiliar to some customers. The menu planner must be careful not to use terminology on the menu that the dining patron does not understand or that is not explained. Generally, if a foodservice customer does not understand the terminology used to describe a menu item, he or she will not ask for an explanation but will simply not order the item. Food can also be wasted and money lost if a patron makes a mistake in ordering a food item through a misunderstanding of the terminology.

Information can be provided on the menu to explain culinary terminology and assist the customer in ordering. For example, the menu can include the terms needed to order the doneness of steaks.

LACK OF LOGICAL ORDER

Food items should be placed on the menu in the order in which they are eaten in a meal. The menu should begin with the appetizer section, continue with the main courses and accompaniments, and conclude with the dessert section. This provides a smooth transition and allows the guest to better and more completely plan his or her menu choices.

TRUTH IN MENU

In writing a menu, it is important to take great care so as to ensure the total accuracy of all information included. To fulfill both ethical requirements and government regulations, management must be careful to avoid inaccurate statements on a menu.

Some states have passed "Truth in Menu" legislation to combat the problem of food service operators misrepresenting the items they sell. These laws are generally enforced by local Better Business Bureaus. Offending restaurants' penalties for violations are twofold: first, fines and court costs, and second, adverse media publicity generated by the violation.

Every statement made orally by a server or written on a menu must be *completely* accurate. The management of a foodservice operation must be able to prove all claims made about the food it offers for sale. For example, fresh-squeezed orange juice must be fresh, not frozen or canned. Ground sirloin must actually be made from sirloin, not another cut of beef. Extreme care must be taken before using descriptions such as *imported, homemade, natural, real,* and *fresh.*

IMPORTANT POINTS TO CONSIDER WHEN PLANNING A MENU

Points of Origin of Ingredients

Menu writers should avoid including the geographical **point of origin** for a food item on a menu unless they are positive that they can obtain a steady supply of that item. Geographical origins of ingredients must be provable by either package labels or invoices. For example, Colorado rainbow trout must come from Colorado, not Idaho. Roquefort salad dressing must be made with Roquefort cheese from the small town in France, not with domestic blue cheese.

Means of Preservation and Method of Preparation

Items called **fresh** on the menu must never have been frozen or canned. Products listed as **homemade** must be prepared on the premises, not purchased already prepared.

Quantity Representation

The size of an item stated on the menu must accurately reflect the actual size of the item served. For example, a "quarter-pound hamburger" must actually weigh four ounces; a "double" cocktail should contain twice the alcohol of a regular drink.

Use of Brand Names

The **brand name** of any product included on the menu must be that of the product actually used. Management cannot mislead a guest by serving an off-brand or a **generic brand** when advertising a name-brand product.

SUMMARY

The menu drives the entire foodservice operation. Foodservice managers must understand the role and purpose of the menu so that they can maximize its potential.

A menu sells an operation's offerings and informs its clientele about what is for sale. It serves as a marketing tool, as a form of advertising, and, with its descriptive copy, as a merchandising tool by enticing the customers to order menu items. A menu also affects the types of food and beverage served, as well as the number and skill levels of an operation's staff.

There are two basic kinds of menus: static and changing, which include five menu types: à la carte, *du jour*, cyclical, table d'hôte, and limited. Each type of menu has benefits and drawbacks, and different types have applications for different kinds of foodservice operations. An alternative used by some operations is a combination of one or more of the various types.

A menu is a powerful selling tool and can be used effectively to influence customers' buying decisions. By offering suggestions on the menu, highlighting special items, and arranging the order of menu items in a logical way, management can affect the sales of menu items.

Menu planners must be very careful that all descriptions accurately reflect the items served. They are bound by ethics and Truth in Menu regulations not to mislead their customers.

KEY TERMS

theme	static menu
concept	changing menu
classical dishes	à la carte menu
market research	*du jour* menu
marketing tool	cyclical menu
merchandising tool	table d'hôte menu
descriptive copy	limited menu
signature item	combination menu

MENU EVALUATION EXERCISE

Learning Objective: Upon completion of the exercise the student will be able to look more objectively at menus. The student will have an opportunity to evaluate a menu according to the criteria discussed in the text.

Assignment: Go to a local foodservice establishment and ask the manager or server for a copy of the menu. Most operators are very willing to distribute copies of their menus.

Answer these questions about the menu you have obtained, and about the restaurant from which it was obtained. Then evaluate the menu according to the checklist that follows.

1. What is the name of the restaurant? Describe briefly the location and type of operation (fast food, cafeteria, fine dining, coffee shop, etc.).

2. What menu type—cyclical, *du jour*, limited, à la carte, table d'hôte, or a combination of types—best describes the menu you obtained? Explain the reasoning for your choice, and give examples from the menu to support your answer.

CHECKLIST FOR MENU EVALUATION

Clientele

1. For what type of clientele is the menu intended? Does the menu appear to cater to that particular clientele by offering menu choices that this group would prefer?

Nutritional Concerns

2. Does the menu provide healthful choices for consumers who are trying to limit their intake of fats and cholesterol? What items on the menu are "heart healthful?" If it does not offer any healthful items, what would you suggest to make the menu more healthful?

Menu Offerings

3. How are the special items on the menu highlighted? Which items does management want to encourage the customers to choose?

Common Menu Mistakes

4. Are any of the menu mistakes discussed in the text found in this menu? What problems do these mistakes cause for the dining customer?

Descriptive Copy

5. a. Do you fully understand the ingredients and method of preparation for all menu items from the descriptions given on the menu? Note any items that are inadequately described and explain the problem this may cause.
   b. Does the menu include any uncommon culinary terminology without clarification? What problems can this cause?

Physical Layout and Condition

6. a. Is the layout of the menu easy to follow and read?
   b. Are menu items listed in the order they are eaten?
   c. Is the menu in good condition, or is it stained or tattered?
   d. In viewing the menu, what is your impression of the operation?

Comments and Impressions

7. What is your overall impression of the menu? What suggestions would you make for improvements?

DISCUSSION AND REVIEW QUESTIONS

1. What are the three goals of a successful restaurant?

2. How can a properly prepared menu help an operation to attain these goals?

3. Name two items of information that a menu provides for guests.

4. How does a menu serve as a marketing tool?

5. Name an advantage and a disadvantage of (a) a static menu and (b) a changing menu.

6. Give an example of a kind of operation that would use each of the menu types discussed in this chapter.

7. Explain how a menu serves as a selling tool.

8. What mistakes are commonly made in producing menus?

9. Why is it important to be truthful in writing a menu?

SUGGESTED READINGS

Kotschevar, L. *Management by Menu* (New York: John Wiley & Sons, 1987).

Scanlon, N. *Marketing by Menu*, 2nd ed. (New York: Van Nostrand Reinhold, 1990).

Seaberg, A. *Menu Design, Merchandising, and Marketing*, 4th ed. (New York: Van Nostrand Reinhold, 1991).

CHAPTER SEVEN

PURCHASING AND RECEIVING

CHAPTER OBJECTIVES

Upon completion of this chapter you should be able to:

- Understand the difference between purchasing and buying.
- State the three spheres or areas within which the purchasing agent must work, as well as their importance and relationship.
- Name two reasons that a purchaser must know how the market works.
- List two advantages and two disadvantages of purchasing goods from a master distributor.
- List two advantages and two disadvantages of purchasing goods from specialty distributors.
- Discuss why some operations purchase some or all of their supplies from a "warehouse club."
- Explain two of the reasons that operations should use purchase specifications.
- Name three items that should be included in purchase specifications.

Purchasing, or the acquisition of goods, is an essential function for any foodservice operation. Goods and services must be "purchased" in order to run the business of serving food and drink. The nature of the foodservice business, with its unpredictable flow of business and the perishable nature of its food products, makes purchasing a difficult task. Operators have to be careful not to order so much food that it will spoil and become unusable, and, at the same time, they must be careful not to order so little that they run out of menu items and disappoint customers.

To be effective, purchasers must be well-versed and competent in three areas: the market, the operation, and the customer. They must know the market—how their suppliers function—so they can best take advantage of it. Purchasers do not buy only items for the kitchen, they also buy items for the various other departments of the operation. Therefore, they must be in close communication with the whole staff to ensure that they buy exactly what is needed. Because the ultimate goal of the operation is to satisfy the customer, the purchaser must be sure to buy things that customers want, at the level of quality they expect, or the customers will take their money and go elsewhere.

The majority of foodservice items are purchased from small producers that do not produce enough goods to warrant a brand name. The lack of a brand name and the consequent possible lack of consistency in products make purchasing more difficult. Specifications, or the specific characteristics of the goods needed, must be developed so that both the buyer and the seller understand exactly what is needed. A large portion of goods purchased, such as fresh produce and meat items, are not manufactured and are thus subject to the variations found in natural items. Purchasers cannot just call their produce supplier and order 10 pounds of apples—they must specify the type of apple, its size, and how they would like the apples to be packaged.

PURCHASING FROM THE CUSTOMER'S POINT OF VIEW

Purchasing is a function that occurs behind the scenes of a foodservice operation and out of view of the customers. The results of the purchasing function—the goods and services available for guest use and consumption—are visible and important to the guests. The quality of the products that guests encounter while dining at the establishment can have a dramatic effect on their perceptions of management's commitment to quality. An operator that uses the least expensive goods available will be seen as a cost cutter and will, accordingly, have problems in charging top dollar for its goods and services.

There are several name-brand food items that customers have grown to recognize and expect. Operators that attempt to save a few dollars by replacing these recognized products with lesser-known goods (which may also be perceived as lesser quality) tend to encounter customer resentment. For example, customers expect a brand such as Coca-Cola or Pepsi—brands with which they are familiar—when they request a cola drink. The substitution of an off-brand will most likely cause problems in the long run.

There is generally a direct relationship between price and quality of goods purchased. Goods at the bottom of the price range are generally inferior to their more expensive counterparts and may eventually cost more for the operation. Thus, operators who look for the cheapest item among the alternatives may be losing money in two ways: through loss of customer goodwill and business and through increased costs resulting from reduced product quality.

THE THREE AREAS OF PURCHASER FAMILIARITY

The role of the purchaser is to find the best-quality product at the best price and to ensure that it arrives at the proper time. In order to do this, a purchaser must be familiar with three key areas: the *market* in which he or she purchases the goods and services for the operation; the *operation*, the establishment for which the purchaser works; and the *customer*, whose needs and expectations must be met (see Figure 7.1).

THE MARKET

Purchasers must be aware of how the **market** functions. Most goods that are purchased go through a chain of distribution from the producer (farmer) to the processor (or a number of processors)

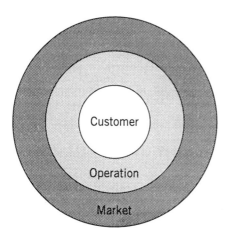

FIGURE 7.1 Three areas with which purchasers must be familiar.

to the distributor (see Figure 7.2). Each business that the product passes through may add value and will most definitely add to the final cost. In purchasing some items, some of the links in the chain of distribution may be bypassed; the purchaser must be aware of the advantages and disadvantages of the various links in order to know which can best be bypassed.

The availability and use of modern and efficient transportation has made it possible for consumers to obtain most food items year-round. The situation today is in sharp contrast with that of 20 to 25 years ago, when some items were available only at certain times of the year and foodservice operations had to develop different menus for the various growing seasons. Although it is available, fresh produce goes in and out of season in various areas of the world throughout the year, causing quality and prices to fluctuate accord-

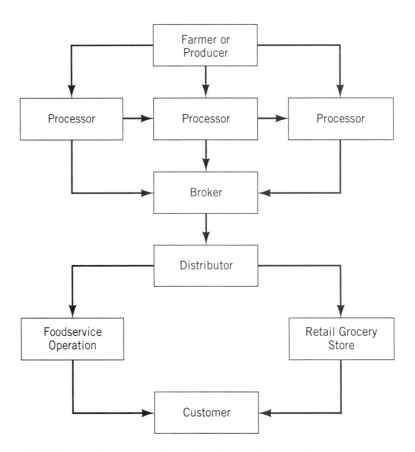

FIGURE 7.2 The chain of food distribution from producer to customer.

ingly. In order to provide the best quality at the best price, the purchaser must know the best time to purchase the fresh produce needed.

Certain areas in this country and around the world produce superior-quality products that can demand a higher price. Some fresh products carry a name brand or are of a certain type that is recognized as exceptional: Indian River citrus from Florida, Driscole strawberries from northern California, and Maine lobsters, to name a few. The purchaser must be aware of these options in order to deliver the best-quality goods for the operation and to determine whether the higher quality is worth the price.

THE OPERATION

The role of the purchaser is to buy the foods and supplies for the various areas of the operation. Therefore, close communication between staff and purchaser is crucial. Purchasers must also be aware of the overall philosophy of the operation. An operation's goal may be to get the best value for the dollar, with quality secondary, or to provide guests with the best that money can buy. In either case, the purchaser's choices of goods should be guided by the goals of management.

Purchasers should work closely with the people for whom they purchase items. A purchaser must avoid judging between two items solely on the basis of price without considering the intended use. When given a choice of products, some purchasers may tend to choose the least expensive. The saying "You get what you pay for" is very appropriate here; an item may be cheaper, but if it is not usable, it represents a total waste of money. For example, a kitchen manager may ask a purchaser to order some tongs. The purchaser sees that he can order "8-inch tongs for $3.75" or "8-inch hot dog serving tongs," which are no more than pieces of bent metal with serrated ends, for $1.25. Even though the hot dog tongs are one-third the price of the other tongs, they are unusable for the task the kitchen manager needs them for and are therefore not a bargain at any price.

THE CUSTOMER

The key to the success of any operation is to meet or exceed the needs and desires of its customers. Customers expect a certain level of quality, which depends on the menu prices and the perceived value of the operation's offerings. A number of products used in foodservice operations are also available for purchase from retail grocery stores. Customers are aware of the quality brands and expect to see some of them when they dine out.

For example, in view of all the various brands of cola drinks displayed on grocery store shelves, did you ever wonder why, when dining out, you generally see only the two leading brands—Pepsi and Coca-Cola? And if an operation serves catsup on the table as a condiment, it will most likely be Heinz catsup rather than any of the numerous other brands that are available from grocery stores. This is not to say that the operation uses only this type of catsup for all applications in the kitchen. Because the customer does not see what catsup is used in the preparation of recipes, a less expensive brand is probably used there. Pepsi, Coca-Cola, and Heinz catsup are not the cheapest brands available nor are they necessarily the best quality, but they are the brands used visibly in most operations because customers equate them with quality and have come to expect them. Purchasers must have a good idea what customers expect so that they can provide accordingly.

WHO DOES THE PURCHASING?

The size of the operation, as well as whether it is an independent operation or part of a chain, can have a dramatic effect on the role and function of the purchaser.

SMALLER OPERATIONS

Smaller operations cannot justify or afford the luxury of having a person whose sole responsibility is the purchasing of goods and services. The task of purchasing is usually delegated to the chef or kitchen manager to be performed along with his or her other duties. An advantage to this arrangement is that because the user of the goods is the person doing the buying, there is no chance for miscommunication. A possible disadvantage is that when the person who places the order is also the one to receive it, a control problem may arise. An unscrupulous person could steal some goods or work with a driver to defraud the foodservice operation. Whenever possible, the person who places an order should not be the person who will receive the goods.

A chef or kitchen manager is generally unable to devote full attention to purchasing because that person already has the responsibilities of his or her primary position. Without being able to devote full time to purchasing, such a purchaser may not be able to obtain the best prices and terms on goods. In delegating the purchasing function to a person as a part-time job along with other responsibilities, foodservice operators generally realize that they may be paying a bit more for things, but not enough to justify the expense of hiring a full-time person to perform this function exclusively.

LARGER OPERATIONS   *Centralization*

Larger foodservice operations, usually within hotels, hospitals, and resorts, have a full-time purchasing agent and staff. The role of the purchasing agent is to deal with suppliers, supervise receiving clerks, and manage the storeroom. Purchasing agents have an opportunity to devote their time to the complete purchasing function for their operations. By managing the storeroom area they can monitor the goods and notify the kitchen staff of any slow-moving foods that must be used before they spoil.

CHAIN OPERATIONS

Operations that are part of a chain generally enjoy the luxury of having some parts of the purchasing function done by the corporate or regional office of the company. The development of specifications and the negotiation of prices and terms with regional suppliers are among the functions with which the corporate office may assist.

Most chain restaurants serve generally the same menu at all of their locations. Because consistency is so important to chains, they usually write up specifications for food items and distribute them to their operations, thus relieving local units of the task. Companies have also found that they can generally obtain better prices and services if they negotiate on a regional or national basis, rather than having each property negotiate on its own for the items it needs.

THE DIFFERENCE BETWEEN PURCHASING AND BUYING

Foodservice operators obtain goods and services for their operation through a formal process, called **purchasing**, or an informal one, considered **buying**. Managers and purchasers must be aware of this distinction and of the respective advantages and disadvantages. Although the net effect for the operation is basically the same—it receives the goods and services it needs—the process of receiving the goods (and most likely the price) can differ. The terms *formal* and *informal* are often used to differentiate the two types: formal purchasing and informal buying.

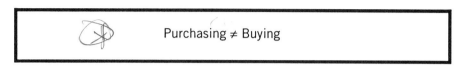

Purchasing ≠ Buying

When properly done, **formal purchasing** is the systematic and planned process of determining what is needed, checking prices, negotiating with suppliers, and obtaining the needed goods (see Figure 7.3). Purchasing involves working with the users to determine which product and which particular form of the product is best. Specifications are written so that the needed goods are consistent from order to order. Once specifications are developed, they are distributed to a number of suppliers in order to find the one that can supply the goods at the best price. It is important that decisions not be based solely on price. Negotiating the price, delivery schedules, and credit terms with the distributor are the responsibility of the purchaser.

Informal buying is much less complex than formal purchasing. The operator, or the person responsible for purchasing, calls one

FIGURE 7.3 Steps in formal purchasing.

FIGURE 7.4 Steps in informal buying.

supplier and orders what the restaurant needs. Specifications are not used; therefore, the quality and yield factors (see Chapter 12) of the goods are left to the discretion of the supplier. Prices are not checked with a number of suppliers, which can cause a problem. A supplier that knows it is not competing with other suppliers could raise its prices without the operator's realizing it, causing the operation to pay more for the goods than necessary.

Formal purchasing is generally used in larger operations that can afford a specialized staff with time to devote to the more complex formal method. Smaller operations generally use the informal system; however, combining the purchasing responsibility with a person's other duties may make it difficult to perform both functions well (see Figure 7.4).

THE PURCHASE SPECIFICATION: KEY TO EFFECTIVE PURCHASING

A purchaser cannot just call the produce supplier and order a case of lettuce. The supplier needs more information—the type of lettuce, the size of the case, the pack of the case (the number of heads), the amount of processing (if any) that the lettuce should have, and the regional origination of the lettuce (if necessary). A purchase specification would include all this information and would make the jobs of the purchaser and the supplier much easier and more efficient. A purchase **specification** is a precise written statement of the product's characteristics required by a user. The random nature of freshly grown produce and the difference in size and quality of animals used for food necessitate written specifications to best serve an operation's needs. The lack of brand names for a majority of foodservice products also makes the written specification desirable.

DEVELOPING AND USING SPECIFICATIONS

Written specifications accomplish the following objectives:

1. Communicate the characteristics of products and services needed for the operation, and help eliminate misunderstandings between the buyer and seller.

2. Provide the receiver with the product characteristics he or she needs to see before accepting goods delivered to the operation.

3. Help to ensure consistency in the items served by the operation by providing consistent products.

4. Allow the operation to solicit bids from more than one supplier.

5. Facilitate the training of purchasers, and allow other people to step in for the purchaser if the need arises.

THE CONTENTS OF A SPECIFICATION

Purchasing specifications (specs) can vary greatly in length. Items with a recognized brand name, such as cans of Pepsi-Cola, can have a specification as simple as the name, size of can, and number of cans in a case. Specifications for items that do not have brand names, such as apples, require much more detail in order to be effective.

The following is a list of some of the more important items that can or should be included when communicating the needs of the operation to the supplier:

1. *The intended use of the product or service.* This is a key element, because it dictates the rest of the information in the specification.

2. *The specific, definitive name of the product.* For example, "apple" is not specific enough, because there are many types of apples.

3. *Packer's or producer's brand name* (if applicable). Most fresh produce and meat do not carry brand names. The purchaser must take care to investigate and include equivalent brand names in the specifications, in case the supplier is out of the specific brand or to increase the number of possible suppliers and thus increase competition in the bidding process.

4. *U.S. quality grades* (if applicable). The U.S. Department of Agriculture has developed a set of quality standards and grades

for a number of produce and meat items. The service is voluntary for producers, but generally allows them to charge more for items that are graded in the upper ranges.

5. *Size or size range of the item needed.* The size of an item can significantly influence the consistency of the product presentation and the yield of the good. If an operator has "costed out" menu prices based on an 8-ounce chicken breast, he or she should be sure always to receive items of that size so that all customers receive the same sized portions and the price printed on the menu remains appropriate.

6. *Package size.* The size of the package has an important impact on the use of the product. The size of a can or package of frozen vegetables can be crucial to an operation; if the package is too small, the operation must open too many packages, wasting labor and probably increasing prices, and a package too large may increase waste for the operation should the food spoil before the operation can use all of its contents.

7. *Preservation and/or processing method.* It is important to include in the specifications whether the product should be frozen, canned, or fresh (fresh means never having been frozen). The state of the product will affect the price and the quality of the item; moreover, the operation must be sure to sell the product exactly as it is offered on the menu. For example, if the menu advertises that the operation serves fresh Rocky Mountain rainbow trout, it would be unacceptable and illegal for it to serve frozen trout.

8. *Point of origin.* Certain areas of this and other countries of the world produce food products that are of exceptional quality. If the operation claims on the menu that it serves Rocky Mountain trout, it must make sure to clarify that particular product in the specifications, because it would be unacceptable to serve trout from any other region and thereby mislead the customer.

9. *Method of packaging.* Does the operation need the product to be bulk packed, or individually wrapped or packaged? The type of packaging affects the operation's utility or convenience; it also usually affects price. For example, when purchasing sliced bacon, the purchaser has the option to buy the bacon sliced and left in slabs for the operator to lay out on a pan, or to buy it sliced and laid out on parchment paper so that all the operator has to do is take it out of the box, place it on a pan, and bake it.

SUPPLIER SELECTION

Purchasers have several options when choosing the type of supplier(s) with which they will deal. Suppliers can either limit themselves to a few types of goods—produce, meat or dairy—or they can be master distributors that provide the majority of the goods needed to run a foodservice business. A number of factors must be considered in making the selection. Within most geographical areas, purchasers have the choice of a number of different distributors in each of the categories of the items they need. The purchaser and the management of the operation must be sure to choose the supplier that best suits the operation's needs. Once suppliers are chosen, it is important to evaluate actual performance against the promises and assurances they have made. Suppliers who do not perform up to the standards of the operation should be replaced with those who will.

CHOICE OF SUPPLIERS

Traditionally, suppliers generally specialized in only one type or perhaps just a few types of items, such as produce, dairy, or seafood, rather than a broad range of goods. The trend in the industry is toward distributors that handle a wide range of both food and nonfood items. In the last five years many companies that distribute foodservice supplies have been consolidated. Some found that it was easier to purchase an existing supplier in an area where they want to expand than to start from scratch. Foodservice operators should be aware of the advantages and disadvantages of doing business with the various types of suppliers so that they can determine which type is best for their operations.

Using a Number of Suppliers or "Specialty Distributors"

The original foodservice suppliers were similar to the original types of retail stores where customers bought their food—these were the **specialty distributors**, who offered only a single type or carried only a limited range of goods. The suppliers specialized in either seafood, dairy, grocery items, or produce. The advantage was that if the company carried only a limited type of goods, it would gain expertise with those goods. It would have a better opportunity to monitor the markets that produce the goods and stay ahead of changes and fluctuations in prices and quality. In a system of this type foodservice operators can take advantage of the expertise of the supplier in the area of the goods that supplier carried. A major problem for the purchaser, however, is that it has to deal with a number of suppliers to obtain all of the goods and services needed. This means that the purchaser has to contact a number of suppliers to place an order and thus has to receive goods from a number of suppliers on delivery days. The more suppliers an operation works with, the more paperwork required.

Warehouse Clubs

Some smaller operators have found it beneficial to purchase goods at local warehouse clubs. These clubs, which are also generally open to the public, allow operators to buy both food and nonfood items in larger packages than those available in retail grocery stores but smaller than packages available from wholesale distributors. The smaller packages, along with the ability to buy broken or partial cases of goods, reduce the small operator's need for storage space and the amount of money tied up in inventory. Operators are usually willing to pay a little more for the goods and to spend a little more time so as to reduce the amount of inventory the business has to carry.

One-Stop Shopping Through a Master Distributor

The trend among foodservice distributors has followed the trend in retail grocery stores—consolidation of many smaller, formerly stand-alone operations into one company, a **master distributor**. Large distributors can offer a wide range of both food and nonfood items. They carry fresh, frozen, and canned produce, fresh and frozen meats and seafoods, and dry grocery items, as well as a full assortment of equipment. A large-scale distributor may carry as many as 10,000 items in inventory, providing the operator with one-stop shopping service.

There are advantages to purchasing the majority of supplies from one distributor. The operator has to make only one phone call to place the order, then receives and puts away goods from only one delivery by one truck, rather than many. Paperwork is drastically reduced because only one check has to be cut to pay the single bill.

The disadvantages of purchasing from a master distributor must also be considered. The salesperson for a master distributor cannot be an expert on all of the many and varied types of products the distributor carries. The operator who relies on a single master distributor usually does not check with other suppliers and may not be getting the most competitive price on the items purchased. A misplaced or incorrect order, or a problem or a delay with a delivery truck, can cause chaos for an operation that relies totally on one supplier.

Type of Supplier	Advantages	Disadvantages
Master Distributor	Increased product offerings	Lack of expertise regarding all of the products sold
	Increased convenience; one phone call, one delivery	Lack of competition
Specialty Distributor	Generally more personal service	Limited product offerings
	Increased competition; can promote better prices and services	Possibility of higher prices because of increased costs and reduction of volume
	Increased product expertise	More time spent ordering, receiving, and doing paperwork

CONSIDERATIONS IN SELECTING SUPPLIERS

Some operators make the error of basing their selection of a supplier solely on price. Operators must realize that, for the most part, all suppliers pay much the same prices to obtain the goods they sell; the prices at which they sell the items are fairly comparable. A significantly lower price on an item quoted from one supplier may mean that the supplier will have to raise prices on other items to make up for it, or that possibly the product is inferior in some way.

The following are factors that purchasers should consider when selecting suppliers:

1. *Credit terms offered*. The length of the period between the time the goods are delivered and when they must be paid for will affect the price of the goods. The longer an operator has use of the goods before paying for them, the greater the expense to the distributor and, generally, the higher the price to the buyer. Operators who can afford to pay for goods at the time of delivery can generally negotiate a better price.

2. *Reputation*. Do not be afraid to ask a supplier for a list of references, names of customers it services. Call the customers and question them about some of the points that are raised in this section. Purchasers should be wary of suppliers who are reluctant to supply a list of people with whom they do business.

3. *Reliability*. Most operators order the items they need close to the time they need them. A truck missing a delivery, or a delivery that is missing items that were ordered, can cause problems for the foodservice operator. The distributor should have a system in place to correct errors or shortages, ensuring that operations get what they need, when they need it.

4. *Substitution policy*. How does the supplier treat a customer when it is out of an item that has been ordered? Does it simply send the invoice with BACK ORDER written next to the missing item, does it contact the customer to suggest a substitute item, or does it simply substitute the house brand or what the supplier thinks is a reasonable alternative? The item that is out of stock may be needed as it comes in for the foodservice operation, and a production problem may arise if it is not delivered when expected. The purchaser should work out a substitution arrangement with the supplier to ensure a smooth flow of goods.

5. *Accuracy*. How closely does the item ordered compare with the item actually delivered? Operators order the amount of goods they need; fluctuations in the amount delivered can cause problems. Short orders may cause an operation to run out of an item earlier than expected; orders that are delivered with an amount greater than that ordered can cause waste and lead to elevated food costs.

6. *Delivery schedule*. Can the supplier send the goods at the time

and date they are needed? Most operations need their deliveries on specific days and times to best suit their needs. The purchaser must be sure that the ordered goods will be received when needed.

7. *Level of technology.* Is the supplier computerized? Does the supplier present neat, readable, computer-generated invoices or scribbled, barely legible slips of paper? Does the supplier allow the purchaser to access its computer directly to place orders, or must the purchaser place orders over the phone?

8. *Lead time.* How much time is required between the time an order is placed and when it will be delivered? Generally, the longer this time span, called **lead time**, the harder it is for the purchaser.

9. *Delivery vehicles and drivers.* The driver is a critical link between the supplier and the foodservice operation. Does the driver take sufficient care with the goods he or she delivers? Does the driver willingly accept the return of material that the receiver feels is unacceptable, or does he or she refuse to take items that the receiver wishes to return? Is the fleet of trucks properly maintained so that the vehicles are reliable and relatively unlikely to break down and thus delay delivery?

10. *Willingness to break cases.* Sometimes an operation does not need to purchase a whole case of a particular item. Dealing with a supplier that allows the purchase of a **broken case** (a partial case) of an item may be convenient for some operators. Smaller-volume foodservice operators should find suppliers that will break cases, in order to reduce their inventory levels and the amount of money tied up in inventory.

BUYER AND SELLER RELATIONSHIPS

There are some aspects in the buyer–seller relationship of which both parties must be aware before entering into a transaction. Some of these matters are governed by laws and regulations, and others are ethical issues.

ETHICAL AND PROFESSIONAL STANDARDS

Purchasers must be careful to adhere to strict ethical and professional standards in performing their jobs. *Ethics* deals with behavior that may not be illegal, but may still be wrong or improper according to commonly accepted practices. All employees who work in a foodservice operation, especially those who deal with suppliers and outside companies, must be made aware of the ethical and professional standards of the company and the industry.

Purchasers must make sure that their relationships with suppliers are kept strictly professional and that all transactions between them are in the best interests of both the operation and the supplier—not for the personal gain of the purchaser.

KICKBACKS *(illegal)*

A **kickback** is the illegal practice whereby a supplier or salesperson pays back money or goods to a purchaser in exchange for an order. Competition for business in some areas is so fierce that some salespeople are willing to pay some of their commission or proceeds from the sale to the person who makes the purchasing decision. This practice is illegal because it gives an unfair advantage to one supplier over another.

LOW-BALL PRICING

Purchasers must also be aware of the practice called **low-ball pricing**. Suppliers may offer lower prices on a few items in order to entice the purchaser to buy from them, and then raise prices on other items. This practice is similar to one used by retail grocery stores that advertise a few items at or below cost as "loss leaders" to draw customers into a store, then hope to make up the lost money with the rest of the groceries the customer buys. The saying "If the price seems too good to be true, it probably is" should guide purchasers' approach to low-ball pricing. Buyers must be aware of the prices of the goods they purchase, as well as the prices on the items sold by competing suppliers, so as to avoid being overcharged in the long run.

CONFLICTS OF INTEREST

Buyers must be sure that they do not benefit personally from their dealings with the operation for which they work. They must inform management if they own any part of any of the businesses that interact with the operation. Purchasers must be careful to avoid the temptation of accepting loans, gifts, items for their personal benefit, or cash in exchange for either giving a supplier an advantage over another company or for simply placing an order. Purchasers should not place their jobs in peril by hiding a **conflict of interest**.

EXAMPLES OF ETHICAL POLICY

The following are examples of precepts that should be included in a policy governing the ethical standards of purchasers:

1. To treat all suppliers equally, and not be swayed by the offering of gifts or other consideration for the personal benefit of the purchaser.
2. To keep the best interests of the company in mind when negotiating all transactions; to conduct all business with others according to company policies.
3. To either limit the value of a gift allowed to be given to a purchaser by a supplier, or establish company policy forbidding the acceptance of all gifts.

RECEIVING

Receiving is the changing of ownership of goods from a supplier (foodservice distributor) to a purchaser (foodservice operator). The transaction normally occurs at the back door or dock of the foodservice operation. The receiver must take care to ensure that all goods listed on the **invoice**, or document that is sent with the food and supplies, are accounted for, and that all quantities, prices, and descriptions of the goods shipped are accurate—in essence, that the operation is receiving all of the items for which it is being charged.

> **KEY POINT IN RECEIVING**
>
> You do not get what you expect, you get what you *INSPECT.*

THE IMPORTANCE OF RECEIVING

A common foodservice truism that signifies the importance of the receiving function is "You do not get what you expect, you get what you inspect." The entire purchasing system can fall apart if a receiver is not careful to check items properly when they arrive from the supplier. Receivers must check three key factors for each item that arrives:

1. *Quantity*. Does the amount listed on the invoice agree with the amount that was actually shipped?

2. *Quality*. Is the quality of the item up to the standards of the operation? Is the product usable?

3. *Adherence to company specifications*. Does the product conform to the specifications drawn up by the operation?

Much time and care go into the formation of the list of items to be ordered. Most operations order only what they need and normally do not carry many extra supplies owing to the perishability of most items and a limited amount of storage space. It is important in the receiving function to ensure that all that was ordered is received.

A distributor charges the foodservice operation for all items on the invoice. If an item that is on the invoice has not been received, the food cost percentage and profit of the operation are adversely affected. Nonreceipt of an ordered item or receipt of a quantity less than that ordered can cause a significant problem if the discrepancy is not spotted at the time of delivery so that it can be rectified.

PROPER TOOLS FOR THE RECEIVER

The person who does the receiving must be supplied with a number of tools in order to perform the job properly.

- *The knowledge* necessary to determine whether a product delivered is within the standards enumerated in the specifications. Specifications are useless unless the receiver can understand the criteria they stipulate.

- *Scales* to weigh the items received to ensure that the weight shown on the invoice is the actual weight of the item being delivered.

- *An adequate area in which to work*. It is important that the receiver have room to open some of the boxes of items received so that he or she can inspect the quality of the items.

- *The specification sheets* of the operation for the products purchased. The receiver should have the specifications handy in case he or she needs to check an item to see whether it is acceptable according to the specifications of the operation.

- *The time* necessary to do a thorough job of checking the items. To do the job properly, the receiver must have adequate time.

- *A list of what was ordered*, which can be checked to ascertain that all items that were ordered have been received and to make sure that the supplier did not send items that were not ordered.

A COMMON RECEIVING MISTAKE

Some managers, when rushed, do not check an incoming order properly and just sign the invoice. This practice can cause a number of problems. Operations tend to build a reputation among delivery people regarding the amount of time they give to checking incoming goods. Truck drivers may tend to make sure that an order is complete and that the quality of goods is at its peak for operations that thoroughly inspect all goods they receive, but they may not be quite so diligent with orders from operations that merely sign invoices without inspecting the goods received. The cost of problems that can result from doing a poor job in receiving an order far outweighs the cost of taking time to do it right.

Many operations mistakenly leave the job of receiving to a dishwasher or bus person. Although this person may do a great job at washing dishes or busing tables, he or she does not necessarily possess the skills required for checking an order. In operations that are not large enough to employ a full-time receiver, the job should be combined with the responsibility of another appropriate position. A kitchen supervisor, cook, or restaurant manager might come in an hour or two early on delivery days to check the order. Once this person is trained in the specifications of the operation and is provided with the tools listed earlier, he or she will be prepared to ensure that the operation receives the goods it ordered and for which it is being charged.

SUMMARY

The functions of purchasing and receiving are crucial to the success of a foodservice operation. The acquisition of food and nonfood products of the proper quality, quantity, price, and at the correct time is essential to the smooth and profitable running of the business.

The purchaser must be aware of the three key areas in order to perform the job properly—the market, the operation, and the customer. The ability of the purchaser to function effectively in these three areas allows him or her to best serve the operation.

The management of a foodservice operation must decide whether it wants to devote the time and effort to institute a formal purchasing program in order to save money on the goods it buys. It also has the option of using a more informal system of buying that requires a reduced investment of time, but this generally means that the operation will pay more for the goods and services it purchases.

The job of purchasing can be done by a full-time person in a larger operation, or it can be combined with the responsibilities of the kitchen manager in a smaller operation. Managers that think the volume of the business does not justify the expense of a full-time employee to handle purchasing should realize that they are paying more money for their purchases.

The management of a foodservice operation must take care to ensure that the operation is receiving the best product at the best price at the time it is needed. Management must also weigh the advantages and disadvantages of the three types of distributors so it can decide which one can best suit the operation's needs. All operation staff involved in purchasing must be made aware of the ethical and professional standards of the position.

KEY TERMS

market	lead time
purchasing	broken case
buying	kickback
formal purchasing	low-ball pricing
informal buying	conflict of interest
specification	receiving
specialty distributor	invoice
master distributor	

DISCUSSION AND REVIEW QUESTIONS

1. What is the role of a purchaser? Why must a purchaser communicate with the other departments of a foodservice operation?

2. What are the three areas within which the purchaser must work? For each, explain why the purchaser must be aware of its function and its needs.

3. How does the purchaser know what goods a manager wants and when he or she wants them?

4. Why do larger operations employ full-time purchasers and smaller operations do not? What are the advantages regarding purchasing for operations that are part of a chain?

5. What is the difference between purchasing and buying? What are the advantages and objectives of each?

6. What are purchase specifications (specs)? Why are they important? Name three essential components of an effective purchase spec.

7. What is the difference between a specialty distributor and master distributor? Name two advantages and two disadvantages of purchasing from each.

8. What would be the advantage of purchasing goods from a warehouse? What types of foodservice operations would be most likely to benefit by purchasing from this source?

9. Discuss the reasoning behind four of the factors that purchasers should consider when choosing supplies.

10. Discuss why buyers should be aware of kickbacks, low-ball pricing, and conflicts of interest.

11. Explain the meaning of the saying "You do not get what you expect, you get what you inspect."

12. What are the three things that suppliers must check when an order arrives?

13. Discuss two of the problems that could arise if management does not take time to check an incoming order adequately.

14. What problems could be caused by allowing a dishwasher to check an order?

SUGGESTED READINGS

Stefanelli, J. *Purchasing, Selection and Procurement for the Hospitality Industry* (New York: John Wiley & Sons, 1992).

Ulm, R. *How Much to Buy* (New York: Macmillan, 1994).

Warfel, M., and M. Cremer. *Purchasing for Foodservice Managers* (Berkeley, CA: McCutchan Publishing Company, 1990).

CHAPTER EIGHT

FOODSERVICE EQUIPMENT

CHAPTER OBJECTIVES

After reading this chapter you should be able to:

- List the six key equipment selection factors and explain the importance of each.
- Discuss the trends in equipment design and how they will affect the use of the equipment in a foodservice operation.
- Name the three primary metals used in foodservice equipment and the characteristics of each.
- List the various types of implements used for commercial mixers and their uses.
- Discuss the advantages and disadvantages of the major pieces of foodservice equipment and the proper uses of each.
- Describe the importance of dishwashing equipment in foodservice operations.
- List the various types of dishwashing equipment, and discuss the operations for which each would be best suited.
- Discuss the steps in the cycle of soiled dishes and tableware, through a multitank conveyor dishwashing machine, from the customer to storage.

This chapter presents an overview of the key points concerning equipment of which a foodservice manager should be aware. The equipment used in a foodservice operation represents a major portion of the opening budget. The significant expense of this equipment makes it necessary for the operator to plan and research each purchase. Selection is very important; errors are costly.

Foodservice establishments require a vast variety of equipment in order to operate. Some foodservice equipment used in restaurants resembles the equipment used in home kitchens; the major differences are the cost, capacity, durability, and ability to be cleaned and sanitized. For example, consider the use of a piece of equipment in a home versus its use in a commercial kitchen: most home kitchens may use a blender for five minutes a day at most, whereas a commercial kitchen may use a blender for a few hours at a time to process the food it serves to its guests. The commercial operation needs equipment that is both larger in capacity and built to withstand the heavier workload.

The size, capacity, and types of equipment a commercial kitchen requires depend on a number of factors. Purchasing foodservice equipment is a difficult task because of the number of choices and the expense of the equipment. Foodservice equipment is purchased when a new operation is being built, when an existing operation is being remodeled to facilitate a change in concept, when there is a meal change, to increase the capacity of an existing piece of equipment, or to replace a piece of equipment that has worn out as the result of years of use or misuse. Each situation requires the purchaser to consider a different set of factors when making a selection. Most selections of new equipment are based on the equipment type, model, or features to which the purchaser is accustomed.

The following sections of this chapter discuss the various features and trends in equipment, selection factors to be considered, and the equipment requirements of the different areas of a foodservice operation. In addition, some key types of foodservice equipment are described.

EQUIPMENT FROM THE CUSTOMERS' POINT OF VIEW

Most of the equipment in a foodservice operation is used out of sight of the guests. With the exception of those things used in the dining room, customers do not see and are not aware of the large array of costly equipment used to produce their meals. This is of benefit to foodservice operators, because they do not have to worry about the visual aesthetics of much of their equipment.

Customers may make assumptions about the equipment in back-of-the-house areas based on the equipment they are allowed to see in front-of-house areas. For example, if their table in the dining room is wobbly, guests may think that because the foodservice operator does not maintain the dining room equipment, the condition of the kitchen equipment is likely to be substandard. In general, the higher cost of better-quality equipment ensures its reliability and durability.

EQUIPMENT SELECTION FACTORS

The substantial cost of equipment used in a foodservice operation makes it extremely important to consider its selection carefully. The following paragraphs describe the factors managers should consider before making a purchase.

THE NEEDS OF THE OPERATION

Equipment lists are drawn up after an operation's menu is developed. Then the operator must decide how much processing will be required for the food items it will create on-site and for those it plans to purchase. For example, will its baked goods be made from scratch using all raw ingredients and requiring a large variety of equipment, or will it buy baked goods already prepared, thereby drastically reducing its need for bakery preparation equipment? Is the operation planning to use a large amount of frozen goods and thus need plenty of freezer space, or does it plan to use foods purchased fresh and therefore need a mimimum of freezer storage space and an increased amount of refrigerator space? The needs of the operation constitute an important factor in determining what equipment to buy.

The menu range dictates the type of amount of equipment needed. Generally, the larger the menu and the more food items available for sale, the more equipment and more types of equipment are needed. The smaller the menu, the less equipment is required. Fast-food restaurants, with their limited menus, require much less equipment than an institutional operation with a larger and more varied cyclical menu.

MEAL PATTERNS OF THE OPERATION

The **meal pattern,** or the rate of customer flow, is another important factor to be considered. Meal patterns dictate both the capacity of the equipment and the volume needed for a time period. Two operations that serve 500 meals for lunch could need equipment of significantly different capacity if one has an even flow of the 500 people over a three-hour meal period and the other operation serves the majority of its meals in the first half of the meal period.

LABOR AVAILABILITY AND COST

The availability and cost of labor is a factor used to determine the feasibility of some types of equipment. Certain kinds of equipment help to increase the productivity of the operation. If the labor market is tight or expensive, it would be beneficial for some operations to purchase food preparation equipment to reduce the need for labor. For example, slicing and dicing attachments for verical cutter-mixers can reduce the need for (and thus the cost of) preparation personnel.

UTILITIES

Before purchasing equipment, the manager must be aware of the types of utilities available. It is generally not feasible for an operator to engage a new type of utility service to facilitate a new piece of equipment. For example, if the kitchen is equipped only with electric appliances, the installation of a gas line for a new piece of equipment would not be cost-effective.

The trend in equipment design has mirrored other trends in the nation. Typically, energy costs have always been high for foodservice operations because of the nature of the business. In many instances it is impractical for food-preparation personnel to shut off preparation equipment between the meals they cook, owing to extended meal periods. For example, most operations that serve lunch are open from 11:00 A.M. to 1:30 P.M.; because guests are likely to come in at any time during that period, equipment must be left on in anticipation of orders. Energy conservation for equipment is currently a major area of development for equipment manufacturers. They hope to reduce the amount of energy that equipment consumes and decrease the warm-up time required.

DESIGN AND FUNCTION

The trend in the design and function of foodservice equipment is toward increased versatility. The high cost of such equipment has encouraged operations to turn away from appliances that perform only one task and toward those that can be adapted to perform a number of tasks. For example, mixers produced by Hobart have a port for various extra attachments including a shredder, a slicer, and a grater. This increased versatility, in effect, allows the operator to get two pieces of equipment for slighty more than the price of one.

Simplicity in both operation and cleaning is important in equipment design, as is a minimum number of parts. Operators also want appliances that require a minimum amount of training for personnel to learn how to operate and clean them. Safety and sanitation are important as well, so operators are looking for equipment without sharp corners, cracks, or crevices where food particles might go unnoticed and where harmful bacteria could breed.

SIZE AND CAPACITY

Examples of equipment sold by capacity are dishwashers (dishes washed per hour), ice machines (pounds of ice produced in a 24-hour period), and mixers (volume capacity of the bowl measured in quarts). It is often difficult to determine the size and capacity of the equipment needed. The problem lies in the fact that an operator does not want to purchase equipment with a capacity too small for the needs of the facility, because it will soon be obsolete. If an operator purchases equipment with a too large capacity, it will take up valuable space in the operation and increase the opening budget, which may subsequently raise operating costs. When determining the size/capacity of the equipment to be purchased, one must look at the immediate needs and the anticipated growth of the business.

Standards of uniformity for equipment sizes are important. For example, steam table inserts, ovens, warmers, and racks are all made by a number of manufacturers. Once standard sizes are determined, operators can interchange pans from the equipment of one manufacturer to that of another, thereby saving money and reducing equipment needs. Manufacturers have worked together to help standardize sizes.

MATERIALS

The material of which a piece of equipment is made should be suitable for the intended purpose. The type of material used influences the price, durability, and sanitation capabilities of the equipment.

The three predominant metals used in foodservice equipment and their primary uses are listed in the following table.

Aluminum	Flexible and generally inexpensive, but changes color with some foods
Cast Iron	Heavy duty, but because of the nature of the metal is not suitable for food-contact surfaces
Galvanized Steel	Used for sinks and hoods, is fairly inexpensive, but is not suitable for food-contact surfaces

CHARACTERISTICS OF FOODSERVICE EQUIPMENT BY AREA

Foodservice operations comprise a number of key areas that work together to process the food it serves. The responsibilities of the various areas differ, as do their equipment needs.

RECEIVING EQUIPMENT

Receiving is the area where the ownership of the goods purchased transfers from the purveyor, or foodservice distributor, to the foodservice operation. Rather than just signing the invoice from the distributor, the foodservice operator must check the quantity and quality of the items received. It is important for the receiving personnel to check the weight of the food indicated on the invoice with what is actually delivered to make sure that the operation is getting all that it paid for. Thus, receiving scales are the most important pieces of equipment in the receiving area. Unfortunately, receiving scales are not used as much as they should be in most operations. The receiving scales should be placed in a convenient location in the receiving area so that it is easy to check the weight of products against invoices from the distributor. Scales are needed in different sizes and capacities, ranging from small food-portion scales to the large floor-model receiving scales.

The receiving areas should also include equipment to help transport supplies from the receiving area to where they are needed throughout the operation. Such equipment increases the productivity of the kitchen by allowing workers to move more items faster with less effort. With multishelved carts, a large amount of food and other supplies can be moved in a minimum amount of trips. Carts are available in many varieties: they can be heated or cooled, designed to hold sheet pans, or open to hold items of various sizes. A two-wheeled or four-wheeled **hand truck** is very effective in moving food and equipment, allowing one person to transport stacks of heavy items with relative ease.

STORAGE EQUIPMENT

Both the volume and the variability of business in foodservice operations require them to buy food and store it for a period of time before using it. For maximum production efficiency, storage areas should be convenient to both the back dock receiving area and the preparation areas. Foodservice operations utilize two types of storage: dry or at room temperature, and chilled, either cooled or frozen.

Dry Storage

Dry, or unrefrigerated, storage is used for items that do not require chilled storage. The predominant equipment in the dry storage area are the racks used to store the food. The racks must be strong enough to support cases of food while providing ventilation. These racks are available in a number of sizes and configurations, allowing operations to customize them to their needs.

Chilled Storage

Refrigeration is considered indispensable for modern foodservice operations, which rely on it for both short-term and long-term storage. Chilled storage provides a number of benefits. The cooler-than-room-temperature condition slows bacterial growth and reduces chances for rancidity. Storage in cooler temperatures also extends the shelf life of fruits and vegetables.

Refrigerators and freezers are basically insulated boxes with a unit that lowers and maintains the interior temperature. A refrigerator has evaporator coils that produce cold air, which is blown by a fan throughout the inside of the box to cool the food. For efficient use, food that is still hot should not be placed in a refrigerator. If warm food must be put into a refrigerator, it should be placed toward the top so that the rising heat does not heat the food on the shelves above.

Refrigeration can be found in a number of forms in most commercial foodservice establishments. The type and size depend on the need and space available in the operation.

Walk-in Refrigerators and Freezers. Walk-in units are the mainstay of most large volume foodservice establishments. A **walk-in**

refrigerator/freezer can be located either inside the kitchen or in an adjacent area with access from the kitchen. A walk-in is big enough to walk into, and its perimeter is lined with storage shelves and places for roll-in carts.

Reach-in Refrigerators and Freezers. These are smaller units that are usually placed closer to production lines and areas that are used for storing food for the meal period. A **reach-in refrigerator/freezer** can be a half-height unit, located beneath a work station or table, or a full-height unit. The insides of the units are lined with either shelves or racks for sheet pans, or are completely open so that rolling racks for food can be rolled in and out as needed (see Figure 8.1).

(a)   (b)

FIGURE 8.1 (a) A full-size double-door reach-in refrigerator. (b) An over-under reach-in refrigerator.

In-Counter Refrigerators.   In-counter units are refrigerated counter wells, generally installed in the pantry or other areas where food must be kept cold prior to service. Food is kept in pans that are submerged into the cold area. An alternative to in-counter refrigeration is to place the food in ice.

Frozen Storage.   Food that is stored below 0°F is considered frozen. Freezers are used for longer-term storage than refrigerators. The further reduction in temperature provided by a freezer slows bacterial action to almost a complete stop and extends the shelf life of a product dramatically.

The term *freezer* is actually not very accurate. A foodservice freezer is designed to store already-frozen food rather than to actually freeze food. The unit requires too much time to bring the temperature of food items to below zero, thus allowing large ice crystals to form, which alters the structure of a food product when it thaws. Foodservice operators should avoid freezing their own food and reserve use of a freezer to store foods that are already frozen.

PREPARATION EQUIPMENT

Preparation equipment is used to help transform food from the raw or semiprocessed state to the ready-to-cook state. This equipment increases the productivity of foodservice workers by allowing them to process a great amount of food more quickly. The need for preparation equipment is directly related to the amount of preprocessed food an operation uses. Some pieces of equipment are not needed by operations that purchase large amounts of processed foods, whereas operations that purchase most of their food in the raw state require a great deal of preparation equipment.

A **mixer** is probably the most multifaceted piece of preparation equipment in the kitchen. Mixers can be used for many more procedures than just combining the batters and doughs of baked goods. With their assortment of implements, mixers can do many things, including beat, whip in air, fold, develop gluten, mix two solids together. Attachments are available to turn a mixer into a shredder, grinder, or slicer (see Figure 8.2).

Mixers come in a variety of types and models: there are counter, bench, and floor models and bowl capacities that vary from 5 to 140 quarts. The smaller sizes are made to sit on a counter or bench, whereas the larger models stand on the floor. The larger models

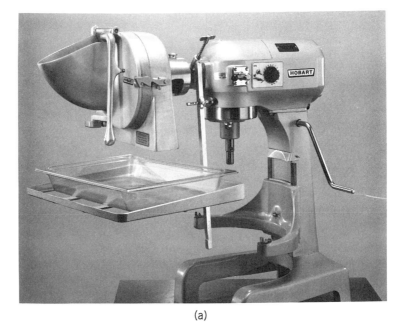

(a)

(b)

FIGURE 8.2 (a) A mixer setup with attachment apparatus. (b) Various attachments available for mixers.

have dollies available to move the bowl around and can be equipped with a motorized accessory to move the bowl up and down (see Figure 8.3).

Commercial mixers generally use just one implement, rather than the two usually found in home mixers. The variety of implement types allows a mixer to do a number of different tasks very effectively. The most commonly used mixing implements available for most commercial mixers are discussed in the following paragraphs (see Figure 8.4).

The **flat paddle/beater** is used for a variety of medium- to

(a)

(b)

FIGURE 8.3 (a) Tabletop mixer. (b) Floor model mixer.

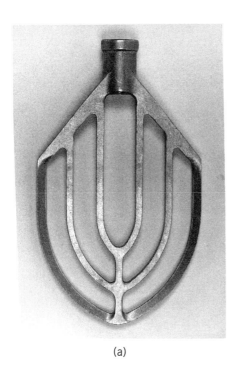
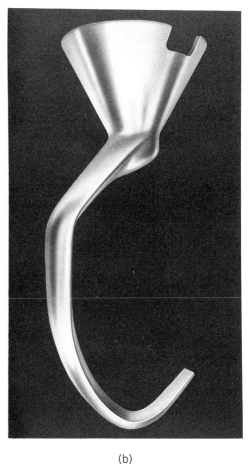

(a)   (b)

FIGURE 8.4 (a) Flat beater. (b) Dough hook.

heavy-duty mixing tasks. It is best used for making batters, mashing potatoes, and creaming butter.

The **wire whip** is composed of a series of heavy-duty looped wires. The whip is used to incorporate air into egg whites and whipped cream and is reserved for light-consistency items.

The **dough hook** (or dough arm) is a large bar that is bent in an offset S-shape. The hook is used for mixing doughs, to help develop the gluten, and for mixing other heavy-consistency items.

The **food chopper** (or food cutter) was one of the first labor-saving devices used in foodservice operations. Food choppers were first developed for the food-processing industry and then adapted for foodservice use. A chopper helps to free the cook from dicing items by hand on a cutting board (see Figure 8.5).

(a)

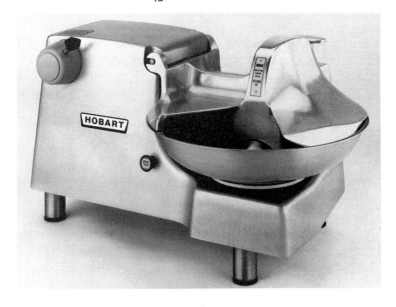

(b)

FIGURE 8.5 (a) Vertical cutter/mixer. (b) Food chopper.

The machine is equipped with a number of safety features to protect the user. Its blades are protected in such a way that the user cannot reach them with his or her fingers while the chopper is operating. A device is also provided that will not allow the motor to run or the blades to turn while the cover is up, further protecting the user.

A food cutter (chopper) comes in various forms and setups and can chop or dice large volumes of food fairly quickly. The machine's shortcomings are that it does not cut items consistently or to uniform size or shape. Although a chopper can process food rapidly, it may not be efficient for small quantities of food because of the time required for setup and cleaning.

There are many instances in foodservice operations when there is a need for accurately and uniformly cut slices of food, such as ham or tomatoes. The **food slicer** is a motor-driven metal disk that is ground sharp on its edge. The food to be cut is placed on a tray that is set on an angle to the blade, so that gravity and a weighted sliding plate pushes the food toward the blade. The thickness of the cutter is adjusted by a dial on the front of the machine. Large-volume operations use slicers with mechanically powered food trays to bring the food to the blade, thus allowing greater productivity (see Figure 8.6).

A slicer is equipped with a number of safety features to protect the user, and other features to make it easier to clean. Handles are provided to allow the operator to keep his or her hands a safe distance from the blade while using the machine. A slicer designed to be easily disassembled and reassembled for cleaning between uses. Some slicers are equipped with a handle that allows the machine to be lifted off the table so that it can be cleaned underneath. A slicer can be mounted on a rolling table so that it can be moved to where it is needed, or it can be remain stationary on a table.

COOKING EQUIPMENT

OVENS

An **oven** is an insulated box with a heat source that is used to cook food. There are a variety of heat sources: simple conduction of heat through the metal lining and racks of the oven, either natural convection or forced convection by a fan, infrared heating, or microwaves. Among the various types of ovens, those that are most common in commerical foodservice are conventional, forced-air convection, microwave, and infrared.

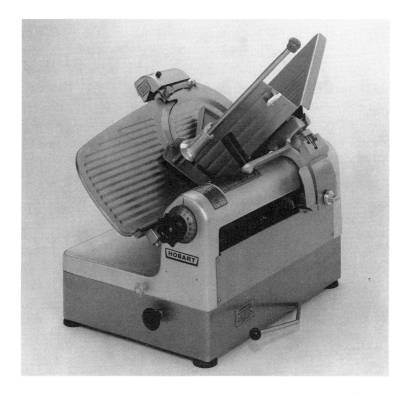

FIGURE 8.6 Food slicer.

Conventional Oven

The **conventional oven** is the most basic form of oven. A heat source is placed in the cavity of an insulated box. Heat moves through the oven using natural convection movement, reflecting the basic principle that hot air rises and cold air falls. This type of oven has been around almost as long as fire has been used for cooking.

The oven sits singularly by itself, is stacked with one or two others, or serves as the base for a range. The only controls are for temperature and, possibly, a timer. This oven is used for general-purpose baking and roasting. Heating elements are either gas, placed below the bottom deck, or electric, which can be placed on the top and the bottom. Some electric ovens have controls so that the user can adjust both the top and bottom heat to arrive at the desired outcome. Heat is transferred by both manual convection and conduction from the oven racks through the pan the food is in (see Figure 8.7).

For energy efficiency, the oven should not be preheated for more

FIGURE 8.7 Double conventional oven with range-top grill and broiler.

than 20 to 30 minutes. In loading, it is best to limit the amount of time the door is open so as to reduce heat loss, and items should be spaced as evenly as possible for even cooking.

ADVANTAGES AND DISADVANTAGES OF A CONVENTIONAL OVEN	
Advantages	Disadvantages
Great for delicate foods	Inconsistent heating
More control with top and bottom controls	More expensive to operate.
	Food may need to be rotated for even cooking.
	Hot and cool spots in the oven

Convection Oven

The **convection oven** was originally developed by the U.S. Navy to increase the efficiency of the standard conventional oven. Because of their higher capacity and more rapid cooking, convection ovens are the main workhorses of high-volume kitchens. A convection oven is usually a stand-alone oven; heat is moved within the oven cavity by a fan. The movement of the hot air results in more rapid cooking at a lower temperature and allows more food to be cooked at one time. The cooking temperature reduction of 50°F and the 25% to 30% reduction in cooking time produced by the rapid movement of hot air also save energy (see Figure 8.8).

ADVANTAGES AND DISADVANTAGES OF A CONVECTION OVEN

Advantages	Disadvantages
Rapid cooking	Greater cooking loss because of shrinkage
Good browing	May not be suitable for delicate foods
Higher cooking production; more food can be cooked at once	
More energy-efficient; food can be cooked at a lower temperature	

Mechanical Oven

Large-volume bakeshops need ovens that can cook a large volume of foods evenly. The **mechanical oven** was developed for that purpose. The difference between a mechanical oven and others is that in a mechanical oven the tray on which the food sits rotates inside the oven on a device similar to a Ferris wheel. While the trays

ADVANTAGES AND DISADVANTAGES OF A MECHANICAL OVEN

Advantages	Disadvantages
Higher capacity	Requires a lot of floor space
Even cooking	Hard to clean

rotate, the pans stay level. The cavity of the oven is heated with either gas or electricity. The rotation of the food in the oven provides for even browning, as well as easy loading and unloading because the trays are always at a convenient height.

Microwave Oven

A **microwave oven** cooks by using short electromagnetic waves to vibrate food molecules, which causes friction and then heat. However, the cavity of this type of oven is not heated. Microwave ovens have limited application for foodservice operations. They are impractical to use for high-production needs because cooking time increases with the amount of food placed in the oven. This type of oven produces a limited amount of microwaves; therefore, the more food in the oven, the less contact each item has with the microwaves, and the slower the cooking. The outside of the food is heated by the microwaves while the inside of the food is heated by conduction. The microwave oven's primary use in commercial foodservice is in reheating small quantities of food.

ADVANTAGES AND DISADVANTAGES OF A MICROWAVE OVEN	
Advantages	**Disadvantages**
Cooks fast	Easily overcooks food
Easy to clean	Does not brown
Does not produce heat	Not suitable for all foods
No fire hazard	

Infrared Oven

The **infrared oven** is used primarily in counter operations and small table-service restaurants that need a fast, general-purpose cooking device that occupies a minimal amount of counter space. This type of oven uses radiant heat, with sources located in the top and bottom of the oven cavity. Infrared ovens are used for reheating individual portions, small pizzas and sandwiches, and items that require a short time in a very hot oven.

ADVANTAGES AND DISADVANTAGES OF AN INFRARED OVEN	
Advantages	Disadvantages
Speed	Takes long to preheat
Does a good job of browning	Burn hazard
Control of both top and bottom heat	Exterior of oven is hot
Occupies very little space	Use large amount of energy

RANGES

A **range** is a multifaceted piece of cooking equipment. The unit can be heated by either gas or electricity, and it is designed to cook food in pots and pans as well as on a flat-topped grill. Further increasing the versatility of this piece of equipment, a range often has an oven and/or a broiler as a base (see Figure 8.9).

FIGURE 8.8 Convection oven.

FIGURE 8.9 Range top over single-oven base.

A range can have any of several different types of tops or a combination of them, depending on the needs of the operation. A top can be flat, made of a thick piece of cast iron to evenly distribute heat. The advantage of this type of top is that it can be loaded with as many pots as will fit, because the entire top is hot. The disadvantage of the flat top is that it takes longer to preheat and it is not very easy to control the heat. Other options are an open-grate top for a gas range and, for an electric range, coils that are individually controlled. The advantages are that these tops do not have to be preheated, which saves energy because they are on only when needed. The disadvantage of the open-grate top is that the number of pans that can be cooking at one time is limited to the number of grates. A third option for a range top is a griddle, a polished metal surface on which food such as pancakes or sandwiches can be cooked directly.

BROILERS

A **broiler** is used to cook individual portions of meat, poultry, or seafood. Broilers do a very good job of cooking steaks, chicken breasts, and some types of fish. However, they are not very well suited to cooking other types of food items that are large and/or require a long cooking time to break down the connective tissue and make the food tender. The broiler is unique in the crusty, almost charred, character that it imparts to food. The heat for the broiler can be generated from either above or below.

There are a number of different types of broilers used in foodservice operations. The heavy-duty broiler is used in high-volume operations that have to broil large amounts of food. The heat comes from the top, and the food is cooked on a grill that slides out. A broiler can be mounted on an oven, a table, or a stand and is typically placed at a convenient level for the cook to monitor the progress of the food that is cooking (see Figure 8.10).

The **salamander,** a light-duty broiler, is normally mounted above a range. A salamander is used to melt cheese, to brown the tops of items, or in other cooking jobs that require top heat. Its convenient location above the range uses dead space and allows the cook to be more efficient (see Figure 8.11).

The **free-standing broiler** is the type most commonly used in table-service restaurants. The heat element, either gas or electric, is placed under a bed of ceramic coals to simulate charcoal cooking. The food is held on a heavy-duty grill that provides the food with distinctive "grill marks" and allows the fat from the product to drain through.

Broilers normally have areas that vary in temperature, owing to the design of the equipment. In a hotel broiler the grill is hotter in the back region, whereas an underfired broiler is hotter in the center and generally cooler around the outside edges. The varying temperatures of the grill are important to cooks. One of the goals of a cook working the broiler station is to have all of the food for a particular table finish cooking at the same time. For instance, in one order there could be a number of different items that cook at different rates: chicken breast, filet mignon, and items that the

FIGURE 8.10 Heavy-duty broiler.

guests order at different degrees of doneness—rare, medium, and well. The varied temperature range of the grill allows different items to cook simultaneously—and to be ready at the same time (see Figure 8.12).

A broiler must be preheated. Once it is turned on, it is normally left on for the entire meal period. Most broilers have multiple side-by-side units so that the cook has the option to only run one-half or one-third of the grill at a time in order to save energy during slow periods. Heat is transferred to the food in two ways—by conduction

FIGURE 8.11 Salamander (top), grill (middle), and single conventional oven (bottom).

from the area on which the food rests on the preheated grill, and by radiation directly from the heat source to the areas of the food that are not in contact with the grill. Because heat comes from only one side, either the top or the bottom, the food must be turned to complete the cooking process. Cooks should avoid turning food with forks that pierce it and cause it to lose some of its juices; long-handled tongs are the proper tool. Once the broiling process is completed, the food should be served as soon as possible.

FIGURE 8.12 Counter model grill on stand.

STEAM-JACKETED KETTLE

In a large-volume operation most nonbaked or roasted foods are produced in a **steam-jacketed kettle.** The advantage of using a steam-jacketed kettle is that the food is cooked over indirect heat. One of the problems with cooking a large pot of food on a stove is that heat is concentrated in a small area in relation to the volume of food, which increases the possibility of burning the food. Another problem is that range space is generally limited and large pots can dominate the stove top, prohibiting other uses of the stove.

The advantage to using a steam-jacketed kettle for cooking large quantities is that the food is heated by hot water or steam rather than by direct heat, which reduces the possibility of burning. Another advantage is that the heat is spread throughout the bottom of the kettle and up the sides, so that the food cooks faster and more evenly because more of it comes in contact with the heated surface.

This kettle is made up of a set of two different-sized bowls that sit inside each other. The outer bowl is sealed on the sides to the inner bowl, and holds the steam that is used to heat it. The cook turns on the steam valve, which fills the cavity between the two bowls with steam. The food to be cooked—liquids and semiliquids work best—is placed in the inner bowl. Kettles size can range from 1 quart to 200 gallons. Each kettle is equipped with a spigot from which the finished product is drained. Smaller kettles, called trunion kettles, can be equipped so that they tilt, which makes them easier to empty and clean. A scraper mixer attachment is available for larger-capacity kettles. The unit is positioned over the kettle and keeps the food moving while it is cooking. This serves two purposes—it speeds the cooking process, and it is much safer than having to stir a 100-gallon kettle of food with a long-handled paddle.

COMPARTMENT STEAMERS

A **compartment steamer** is energy-efficient and versatile. Such vessels are able to cook items without the addition of fat and ensure that more of the product's nutrients are preserved throughout the cooking process. The moist heat it provides makes the steamer ideal for reheating leftovers, as compared with the dry heat of an oven that tends to dry food that is being reheated.

Food to be cooked in a steamer is placed in perforated pans to allow the steam to penetrate better. The pan is placed in the steamer, the door is closed and latched, the steamer is turned on, and the compartment is filled with steam. Steam, or condensed

ADVANTAGES AND DISADVANTAGES OF A COMPARTMENT STEAMER	
Advantages	Disadvantages
Fast	If not set properly, can overcook
Preserves color, texture, and moisture	Does not brown
Retains a large percent of the foods nutrients	Cannot be used for items with a crust
Energy-efficient	

water, which is much hotter than boiling water, cooks food faster than other processes.

Steamers are ideal for cooking fresh or frozen vegetables. Poultry and seafood dishes that are normally poached work well in the steamer. An alternative to boiling eggs in water is to steam them. Chicken to be fried can be partially cooked in a steamer, then cooled and breaded to reduce frying time. Using a steamer requires care; its high heat and rapid cooking can cause overcooking. However, a steamer's quick cooking allows operations to cook food in small batches as it is needed, rather than having to cook large batches and hold them until they are to be served.

DISHWASHING EQUIPMENT

A dishwashing machine is the most expensive piece of equipment, both to purchase and to operate, in a foodservice operation. In most operations the dishwashing machine is operated by the lowest-paid people on staff. It is also one of the most important pieces of equipment from both the customer's and the board of health's point of view, because of its crucial role in cleaning and sanitizing tableware. Customers do not want to eat from dirty plates or drink from dirty glasses. The health department insists that the service items customers have eaten from and/or have put in their mouths are properly cleaned and sanitized before others use them.

Prior to the emergence of mechanical dishwashing equipment, foodservice dishes were washed by hand. Washing large volumes of tableware manually presented a number of problems. The soap concentration and water temperature were difficult to monitor. Most commercial operations now use mechanical dishwashing machines. Some small and seasonal operations that still wash their dishes manually are closely regulated by the health department.

All managers in a foodservice operation should know how to set up, operate, and perform simple maintenance on the dishwashing machine, as well as break down, or disassemble, the machine at the end of the day. A steady supply of clean and sanitary dishes is crucial to the running of a foodservice operation. Washing the operation's dishes is one of the most important jobs, yet it is usually left to the lowest-paid person on the payroll. If for some reason an operation is without a dish washer, the management staff should be able to operate the machine until a replacement can be called in. It is imperative that the operation have a service contract on the dishwashing machine to keep the down time of the machine to a minimum. Simple maintenance should be performed by management until service personnel can arrive and fix the machine.

A manager has a choice of several types of dishwashing machines. The following paragraphs discuss the different types of machines, how each works, and its best uses.

Single Tank, Stationary Rack

In using a single tank, stationary rack dish machine, dishes and glasses are first sprayed with hot water and scraped. They are then placed on a rack, and the rack is placed in the machine and it is turned on. Wash arms above and below the rack spray the dishes and glasses with a high-pressure spray of hot water and detergent. Rinsing and sanitizing take place after the washing is completed. The dishwater operator then removes the rack of dishes from the machine and allows them to air dry on a counter. It is very important that the items that come from the machine be allowed to air dry, rather than risking the possibility of recontaminating them by towel drying. This is the most basic type of dishwashing machine and is generally used in relatively low-volume operations.

Rack Conveyor

As with the use of a stationary rack, dishes and glasses are sprayed and scraped prior to being placed on the racks of a rack conveyor machine. The racks are then set on a moving conveyor that runs them through the machine. The washer can have either a single tank, as in the previously described machine, in which all three activities, washing, rinsing, and sanitizing, are accomplished in the same tank, or it can use dual tanks, which include a prerinse prior to the wash cycle.

The rack conveyor is able to handle more dishes within a period of time than the stationary rack machine. This is important, because the use of dishes is concentrated at meal periods and dishes must be reused during a meal. The drawback is that two people are necessary to operate this machine efficiently—one to load, and the other to catch the items at the receiving end.

Belt Conveyor or Flight-Type

A belt conveyer flight-type machine has many of the characteristics of the rack conveyor machine. It differs from the rack conveyor in that dishes are placed directly on a belt with pegs that runs through the machine, rather than on a rack. Glasses, cups, and some types of bowls are still run through on racks for protection of the items and for the sake of efficiency. Glasses and cups are generally left in these racks for storage in the kitchen and dining room.

Carousel Type

The carousel-type machine can be either a belt conveyor or a rack conveyor system. The basic difference between the carousel type and the belt conveyor is that the carousel conveyor makes a continuous loop around the machine. The advantage of the continuous loop is that one person can load the machine, wait until the items come through, then unload the machine. This allows the operation to reduce labor requirements during off-peak periods by having the same person load and unload the machine, although it is not recommended because of the possibility of the person contaminating the clean, sanitized dishes.

Chemical Sanitizing Machines

The chemical sanitizing machine was developed to reduce the need for the expensive hot water that other types of machines use. As compared with the other types of machines discussed, chemical sanitizing machines operate using lower temperatures, higher water pressure, and increased amounts of chemicals. Such a machine relies on chemicals, usually chlorine, rather than hot water to reduce the bacteria count on tableware to a safe level. A reduction in cost realized by the use of less hot water is partially offset by increased chemical cost.

SUMMARY OF STEPS IN A MULTITANK CONVEYOR MACHINE

All foodservice tableware passes through a cycle, from use by guests to storage in the kitchen. The following list summarizes the steps in a cycle that uses a multitank dishwashing machine.

1. *Scraping.* Soiled dishes are removed from the table and brought into the dish room. Any remaining food is scraped off and the flatware is placed in a soaking solution.
2. *Prerinsing.* The dishes are prerinsed to remove all visible soil. (Food particles that remain on dishes as they go through the

dish machine reduce the effectiveness of the detergent and can clog up the filters of the machine.)

3. *Racking up.* The dishes are placed in the rack evenly so that the water spray from above and below can contact all surfaces. They can also be prerinsed after they are racked.

4. *Washing.* The rack of dishes is placed on the conveyor and is taken into the wash cycle of the machine. In the wash cycle the dishes are hit with a 140°F to 160°F water and detergent solution from above and below.

5. *First rinsing.* Once the wash cycle is complete, the dishes are sprayed with 160°F rinse water. The water is sprayed over the dishes from above and below, as in the wash cycle. The purpose of the rinse is to remove any remaining soil and detergent.

6. *Final rinsing.* The rinse cycle leads to the final rinse portion of the washing cycle. The water should be 180°F and freshly pumped from the water heater.

7. *Sanitizing.* The final step of the dishwashing cycle is sanitizing, which is necessary to reduce harmful bacteria to a safe level. It is accomplished by heating the surface of the dishes to at least 160°F, the temperature at which bacteria cannot survive. Some machines use a chemical sanitizer along with heat.

8. *Drying.* It is imperative that dishes be allowed to air dry once they leave the dish machine. Towel drying has the potential to spread bacteria to the sanitized plates. Fortunately, the high temperature of water used to wash and sanitize the dishes speeds up the air-drying process.

9. *Stacking.* In stacking clean and sanitary dishes, care must be taken that the dishes, as they exit the machine, are not handled on surfaces that food or customers' mouths will touch. It is also important that the person who scrapes the plates and loads the machine washes his or her hands prior to unloading the machine so as not to contaminate the sanitized dishes.

10. *Storing.* The dishes must be stored in an area where they will not be contaminated.

SUMMARY

Foodservice equipment aids workers to be more productive and to perform their jobs more efficiently. The purchase of equipment in a foodservice operation represents a significant part of the opening budget. The selection and purchase of foodservice equipment is an important function of managers. The number of factors that must be considered—the expense and longevity, the various types and options available—complicate the process. Foodservice managers must be aware of the selection factors to be considered, as well as the trends and the characteristics of the various types of equipment used in each area of the foodservice operation, in order to better and more effectively manage their operations.

KEY TERMS

meal pattern
standards of uniformity
hand truck
walk-in refrigerator/freezer
reach-in refrigerator/freezer
frozen storage
mixer
flat paddle/beater
wire whip
dough hook
food chopper
food slicer
oven
conventional oven
convection oven
mechanical oven

microwave oven
infrared oven
range
broiler
salamander
steam-jacketed kettle
free-standing broiler
single tank, stationary rack
   dish machine
compartment steamer
rack conveyor machine
belt conveyor/flight type
   machine
carousel type machine
chemical sanitizing machine

DISCUSSION AND REVIEW QUESTIONS

1. How does the meal pattern of a foodservice operation affect the selection of equipment?

2. Name two trends in foodservice equipment design.

3. Name three attachments for a food mixer and explain what each is used for.

4. Why is the term *freezer* inaccurate for that piece of storage equipment? What is the proper use of a freezer?

5. What is the difference between a conventional and a convection oven? Name one advantage and one disadvantage of each.

6. Name the three options for a range top. What factors might determine a foodservice operation's choice among them?

7. What is a salamander and what is it used for in a kitchen?

8. Which piece of foodservice equipment is the most expensive both to purchase and to operate? Why should management have the ability to operate and maintain this piece of equipment?

SUGGESTED READINGS

Frable, Jr., Foster. *Equipment and Facilities,* various columns in *Nation's Restaurant News,* the news weekly of the foodservice industry.

Scriven, C. and J. Stevens. *Food Equipment Facts* (New York: Van Nostrand Reinhold, 1989).

CHAPTER NINE

FOOD PRODUCTION

CHAPTER OBJECTIVES

Upon completion of this chapter you should be able to:

- Describe the goal of production for a foodservice operation.
- Discuss the obstacles to the goal of production that are inherent in a foodservice operation.
- State the possible solutions to these obstacles.
- Discuss the advantage of stagger cooking menu items.
- Name the eight steps of production through which food items pass in a foodservice operation.
- Describe the three ways to measure food items.

Production is the culmination of efforts in all other areas of a foodservice operation. The food products that are served have passed through many areas of the operation in their journey from the supplier to the plate served to the customer. At each stage of the journey it is important to ensure that each product is handled properly and that its quality meets the standards of the operation.

Control of the production function is essential to the success of any foodservice operation. Over- and underproduction of food items will cost an operation both customers and reduced profits. Operations that are constantly running out of items because of poor production planning will drive customers to the competition. Kitchens that overproduce food items, without the ability to serve those items, will suffer from increased food costs and a reduction in profits. Kitchen management should take an active role in planning how much of each menu item to prepare.

A foodservice manager must understand how the production area of the operation works. Although a manager does not need to be trained specifically as a cook, he or she should understand the importance of the production area and how it affects the operation of the entire foodservice establishment.

This chapter discusses the various stages through which food passes, from the supplier to serving the guest. Each area of the operation must do its job in order for the guest to receive a quality product.

PRODUCTION FROM THE CUSTOMER'S POINT OF VIEW

Customers of a foodservice operation do not generally see the production, or the preparation, of the food they eat. Most of the production goes on behind the closed doors of the kitchen, except in those operations that have open or demonstration kitchens where customers are allowed to see their food being cooked.

Even if customers do not see the actual production or preparation of the food items, they do see and eat the final products. They are the final and most critical judges of the work of all areas of the foodservice operation that contributed to the effort. If presented with a product that does not meet his or her standards, the customer rightfully blames the restaurant for not doing a good job.

THE GOAL OF PRODUCTION

> The goal of production should be the goal of the operation: to serve the customer the highest-quality product posssible in a reasonable amount of time.

This goal sounds simpler than it actually is. To accomplish it, as mentioned earlier, all areas of an operation must work together. One breakdown in the line of communication can disrupt the flow of goods through the operation and cause a problem for the guest, thus reducing the guest's enjoyment and defeating the goal.

AN OBSTACLE TO THE GOAL OF PRODUCTION

In the preparation of most menu items there are multiple steps that require varying lengths of production time. Customers who eat in a foodservice facility have a limited amount of time to eat, and if all steps in the recipes were left to the last minute, customers would not be served within the meal periods available to them.

A quick solution to this obstacle to the goal of production is to produce the food ahead of time, so that all that has to be done when an item is ordered is to serve it to the guest. As simple as this sounds, it contradicts the goal of providing the highest-quality food possible. Most foods deteriorate in quality immediately after they are prepared and as they are held for service. Therefore, although preparing food ahead of time will speed up service to the customer, the quality of the food may be inferior.

POSSIBLE SOLUTIONS

To solve the problem posed by customers' time limits, as well as to satisfy the two-part goal of production, extensive planning is necessary. The kitchen manager or chef must examine the following for each menu item:

1. *Steps of Production.* The recipe should be separated into the various steps of production to determine which steps can be done ahead of time without causing a reduction in the quality of the item. Recipes are made up of a series of steps that are followed in order to produce the various food items. Some steps, such as compiling the ingredients, cleaning, peeling, and portioning, can be done in advance. Some items can also be partially precooked prior to service and held cold without a reduction in quality. For example, sautéed fresh sliced carrots served as an accompaniment to the meal can be pre-prepared in several steps prior to service. The carrots can be cleaned, peeled, and sliced, partially cooked or blanched, and then cooled. When they are needed, they are simply reheated to finish the cooking and seasoned. The quality of the dish does not suffer from such pre-preparation.

2. *Final Cooking.* The final cooking of an item should be done as closely to the time needed as possible. The major reduction in quality of most items occurs once they are heated and held hot, waiting to be served. If as many steps as possible are left for the time of service, the goal will be satisfied: the customer will be provided with a high-quality product within a reasonable period of time.

For items that must be cooked ahead and held hot, **stagger-cooking** of the item is recommended. Vegetables, roasts, and baked potatoes are all items that take time to cook, but whose quality can be improved by staggering the time they are cooked throughout the meal period. Prime ribs can be cooked in such a way that they come out of the oven several times during the dinner period, allowing the operation to sell medium-rare cuts through closing time and reducing the amount of time the ribs must spend in the warmer.

Baked potatoes, for example, must be cooked ahead and held hot, but while being held, their quality is reduced. A way to minimize this problem is to stagger-cook them. If the dinner shift runs for four hours from five to nine o'clock, there is no need to cook all the potatoes at once, to come out of the oven at the same time and be held the entire dinner period. Stagger-cooking them, or forecasting the need by the hour and planning to have a fresh batch come out of the oven each hour, ensures that each customer gets a fresh, hot potato. Stagger-cooking takes more time by the preparation personnel but provides the customer with a quality product.

3. *How to Hold Items.* Determine the best way to hold items in the various stages of production. Some are best held frozen, such as french fries, so that they do not absorb too much grease when deep fried. Other items, such as peeled potatoes, should be held under water to avoid oxidation or browning while being held. Food production personnel must be aware of the best holding methods for the food they serve.

4. *Reuse of Menu Items.* Kitchen staff should plan for the reuse of menu items that have been made ahead and are not served. Although a food item may be reduced in quality in its present state, it may be possible to process it to be used in another form so as to reduce waste. Baked potatoes that have been held in the warmer too long to be served for their original use can be used for potato skins or hash brown potatoes.

5. *Priority: Quality.* Remember the goal of the operation. The priority to maintain quality should be communicated to all employees. If a menu item cannot be adapted so that it retains its high quality and can be served in a reasonable amount of time, it should be replaced on the menu by items that can meet these criteria. Customers remember the quality of the food they are served, as well as the timeliness with which it is served.

THE STEPS OF PRODUCTION

As mentioned earlier, a food product goes through a number of steps from the time it is received from the supplier until it is served to the guests. This process combines a number of interdependent operations that must be coordinated in order to serve guests. The size of the operation and the nature of the food it serves determine whether the various steps are done by the same person, or department, of the operation.

Steps in Food Production

1. Ordering	5. Handling
2. Receiving	6. Preparing
3. Storage	7. Portioning
4. Measuring	8. Serving

ORDERING

Ordering, or purchasing, is discussed in greater detail in Chapter 7. Purchasing, the process of arranging for delivery of the items needed to run a business, is the first step in food production. Production cannot start or continue without products delivered to the production staff. The quality of menu items served to guests is directly related to the quality of the products purchased. Even the most skilled chef cannot overcome the negative effect of poor-quality or inferior products. In the ordering and purchasing of goods it is important to ensure that those goods are of the quality that the customer expects.

RECEIVING

Receiving is the process whereby the ownership of the food purchased transfers from the distributor or seller to the buyer (the foodservice operation). The receiving function also has a dramatic effect on the quality of the final product that is served to the customer. The time and effort put into the purchasing function can be in vain if an operation is not careful to ensure that the quality delivered is the same as the quality purchased. As noted in the section "Receiving" in Chapter 7, in regard to the receiving function, *"You do not get what you expect, you get what you inspect."* The staff who do the receiving must be sufficiently knowledgeable and have the necessary skills to determine the quality of food. For example, they must be able to determine whether the goods are of the quality grade ordered and meet the product specifications of the operation.

STORAGE

All food items delivered must be held in storage for a period of time between the time it is received and when it is served. It is impractical to have food delivered for each meal period, so it is purchased and held until needed. Most food products generally deteriorate in quality while in storage, while also tying up the capital of the operation. The current trend in the design of foodservice operations is to reduce the size of storage areas so as to decrease the amount of goods held in storage and encourage the turnover of the inventory.

It is important to make certain that food in storage is being rotated. The common practice of inventory rotation is a system called First In, First Out (FIFO). **FIFO** helps to ensure that the oldest products are used first. Inventory items can be dated and new items placed behind old items to encourage production personnel to use the oldest products first.

MEASURING

Foodservice items are normally purchased in large quantities. The purchasing of goods in larger quantities than are appropriate for home use helps to lower the per-unit cost as compared with the cost of the same items as a retail grocery outlet. Food items should be taken from large packages and measured to the size required by the recipes to serve the operation's guests.

Measuring food items, by weight or by volume, is an important step in production. Measuring by weight, using scales, is more accurate than measuring by volume. Generally, dry goods are listed in recipes by weight, whereas liquid ingredients are listed by volume. Production personnel should be provided with accurate measuring tools in order to do this task. Employees who cannot readily find the measuring devices needed for the recipe they are preparing will tend to guess on the measurments, thus throwing off the accuracy of the recipe, the consistency of the final product, and the per-unit cost.

HANDLING

Once the food items are measured, the next step is to get them ready for preparation. The handling step involves such things as washing, cleaning, peeling, slicing, dicing, and cutting. These can all be done prior to the time the food is need for cooking or combining with other food items.

Foodservice operators have several options for accomplishing the tasks in this step of production. In operations that are short of help, produce items, for example, can be purchased with many of these tasks already accomplished. A food item purchased in this state is more expensive; management must balance the extra cost with the reduction in the cost of labor. Meat items can be purchased in ready-to-serve portions, thus reducing the amount of handling the foodservice operation must perform. The quality of pre-processed or pre-handled food has increased dramatically in the last few years. In the past, food available in this category was limited to canned or frozen foods. Today a vast number of choices are available in most segments of the food industry.

Another option for management, especially viable for large-volume operations, is to purchase equipment that can perform some or all of the handling tasks. Using equipment is much quicker than doing most things by hand, but the time necessary to set up, break down, and clean a machine must be factored in. For smaller quantities of items it may be more time-efficient for an employee to perform the necessary tasks by hand, rather than take the time to use a machine.

A third option is for the foodservice operation to use lower-paid cook assistants to perform the tasks in the handling step and save the higher-paid cooks for cooking. Why pay a $7.00-per-hour cook to weigh and peel carrots, when a $5.00-per-hour cook's assistant can do the same thing, thus freeing the cook to actually "cook" the food instead of handling it? Some operations assign dish washers and pot washers to do the peeling and cleaning of vegetables. The work of both groups of workers is normally cyclical, based on the meal periods. The slack time between meals is the ideal time for them to peel vegetables or crack eggs for breakfast. This arrangement requires kitchen management to break down the recipes and prep lists into separate tasks for each category of worker. The time and effort expended to do this can quickly be made up in greater productivity and reduced labor costs.

Whatever option management chooses, it must be that which best fits the goals and objectives of the operation. Different types of operations have different needs and function in different labor markets. The quality of the product served to the guest should be the priority. An operation may be able to save money on labor by using a machine to dice carrots, but if the quality is inferior, it will result in a loss of customers and, despite the cost savings, the operation will lose money.

PREPARING

Preparing is a step that requires the most skilled staff and has the greatest impact on the final quality of the products served to guests. The tranformation of food items into menu items is the ultimate purpose of the kitchen operation.

The planning of the menu, along with the types and amounts of equipment used in the operation, are the keys to the preparation phase. The food items that require the use of various pieces of equipment, such as the broiler and fryer, must be considered carefully to ensure that no single area or piece of equipment in the kitchen is overloaded, possibly causing a bottleneck or slowdown of in production. The goal is to have all items for a group of people eating at the same table ready at the same time. This requires the skill of the kitchen personnel to know which items take what amount of time and to schedule them so as to ensure that all food finishes cooking at the same time.

Different items require different amounts of preparation time. Some items cook quickly or are already cooked and require a minimal amount of time to serve. Other items require several steps and therefore require a longer cooking time. Balancing the cooking time of items in both groups is essential to a smooth-running operation.

Preparation is the area of the operation that has the greatest impact on the quality of the food. Management must be sure to hire cooks with the skills necessary to produce the desired menu items. Customers generally prefer simpler menu items prepared properly to more complex foods prepared improperly. A common foodservice problem arises when operations include items on the menu that the cooks cannot cook properly.

PORTIONING

Portioning is the final step in the food preparation process before the food is served to guests. This is one of the key components of the cost-control system in an operation. If the price of a shrimp cocktail appetizer is determined using five shrimp and the pantry person uses six by mistake, the foodservice operation has most likely lost money on the item. Management must make sure that employees are aware of the correct portion sizes.

Items can be portioned in several ways: by count or number, by weight, and by volume.

1. *By count.* Some items, by their nature, are portioned by count. These items are generally sold by weight or count, such as shrimp, which is sold by the number of shrimp (within a range) per pound, and eggs, which are sold by size and portioned by count. This makes portioning easy.

2. *By weight.* Most food items are portioned by weight. The weighing of items is time-consuming and can be inaccurate. Fortunately, many items can be portioned prior to service time, thus speeding the process and increasing accuracy.

3. *By volume.* Some items are sold and portioned by volume. There are several tools used in this method of portioning. Cups and bowls are sold by the size or number of ounces—for instance, a 10-ounce bowl, an 8-ounce cup. Because the cook simply fills the serving container rather than having to measure, the process is speeded up. Ladles, used to portion liquid items, are sold in a variety of sizes and are basically measured bowls with long handles. Portion scoops, commonly called ice cream scoops, are used to portion much more than ice cream. They are quite effective in portioning semisolid food items such as mashed potatoes and chicken or tuna salad. These scoops come in a variety of sizes and are marked inside with the number of scoops per quart.

SERVING

The link between the production area and the customer is the service area of the foodservice operation. Serving is the last step in the production process and is crucial to the customer's dining enjoyment. The guests' impressions of the foodservice operation may be determined by the server and the quality of service the operation provides. An operation that spares no expense in hiring kitchen personnel and purchasing food products but relies on subminimum-wage servers to deliver food to its guests should rethink its allocations of funds.

Servers must be trained in the methods of accurate and effective service, as well as the basics of the menu. Coordination and communication must be developed and maintained between the kitchen and serving areas. Many operations suffer from a rift between the front-of-the-house (dining room staff) and the back-of-the-house (kitchen staff), which is destructive to their smooth functioning. Front-of-the-house and back-of-the-house staff must work together to take care of guests. Some establishments allow employees from these areas to switch places during slower shifts so that each group can better understand the other's job.

SUMMARY

Although customers generally do not see much of the actual production of a foodservice operation, they do see the products of production. The food products that foodservice operations serve to their guests represent the culmination of a number of steps or processes within the operation. All areas of the foodservice operation, from ordering to service, must perform well in order for guests to receive a quality product.

The nature of the foodservice industry, coupled with the properties of the various food items, complicate the matter of production. Most areas of an operation have to be ready for the arrival of customers during an extended period of time, or during a specific meal period. Food items must be available to serve within a reasonable amount of time after they are requested by the guests. The preparation of food items typically requires too many steps to wait until an item is ordered to begin its preparation; yet preparing items in advance can be problematic because most deteriorate in quality once they are prepared. Careful planning by management is necessary to ensure that as many steps of production as possible are done in advance without causing a drastic reduction in the quality of food items.

Food goes through a number of steps in its journey from the producer to the customer. Each step is critical in meeting the goal of the foodservice operation—satisfying the guests. If any of the areas of a foodservice operation fail to perform adequately, the quality of the product served to the guests will suffer and, possibly, the guests will be dissatisfied. Remember, most guests do not complain, they just leave the operation and report negatively to their friends. Care must be taken in each step of production to ensure that guests receive quality products.

KEY TERMS

production	storage
goal of production	FIFO
obstacle to the goal of	measuring
production	handling
stagger-cooking	preparing
ordering	portioning
receiving	serving

DISCUSSION AND REVIEW QUESTIONS

1. Why must foodservice operators know the basics of production? Does a person have to have been a cook prior to being a manager? Explain.

2. What is the goal of production for a foodservice operation? Discuss an obstacle to this goal and how it is best solved.

3. What is stagger-cooking, and how can it solve problems in production? Name three items that can be stagger-cooked without a reduction in quality.

4. A newly hired manager of a table-service restaurant encounters the problem that some menu items come up from the kitchen quickly; others are very slow. What is the probable cause, and how can the problem be solved?

5. Why is it important for foodservice managers to understand all aspects of the steps of production. How is the customer affected if one or more of the steps fails in its function?

6. What is FIFO? Why is it important? What can be done to ensure that this procedure is followed?

7. Name two ways to measure food items. Which is more accurate?

8. What are the three ways to portion food items? Name two items that would be best suited for each method of portioning. Explain the choices.

9. Why is portioning a key component of a foodservice operation's cost-control system? Why is it crucial for a foodservice operator to be concerned with cost control?

SUGGESTED READINGS

Gisslen, Wayne. *Professional Cooking*, 3rd ed. (New York: John Wiley & Sons, 1995).

Labensky, S., and A. Hause. *On Cooking, A Textbook of Culinary Fundamentals* (Englewood Cliffs, NJ: Prentice-Hall, 1995).

Pauli, E. *Classical Cooking the Modern Way* (New York: Van Nostrand Reinhold, 1979).

CHAPTER TEN

SERVICE AND DINING ROOM MANAGEMENT

CHAPTER OBJECTIVES

Upon completion of this chapter you should be able to:

- State the importance of service to the customer.
- Describe two ways in which service is important to guests.
- Tell how service determines value for the dining customer.
- Discuss how customers judge servers and what customers expect of servers.
- Describe the role of servers in the foodservice operation.
- Distinguish between the various categories and types of service.
- Compare the advantages and disadvantages of formal and informal methods of service.
- Describe the obstacles to good service.
- Be aware of the importance of proper training for servers.
- Consider the suggestions from restaurateurs on how to improve service.

The purpose of this chapter is to stress the importance of service to the customer's dining experience. It discusses the server's responsibilities and presents some tips for servers on how to increase the level of service and enhance the guests' enjoyment. Suggestions are provided on how to implement a training program to improve service. Some service problems are also discussed so that future managers can be aware of the obstacles to providing good service for their guests.

There are several methods or styles of service used in dining rooms around the world. This chapter is limited to styles of service used in this country. French, English, and American service are presented, as well as the different types of self-service.

There has been a great growth in American cuisine over the last decade. The emergence of the American chef has caused excitement and pride in the country's restaurant community; however, the level and quality of service have not kept pace with these developments. The primary contention of this chapter is that it is not difficult for a foodservice operation to provide good service. Customers appreciate attention and commitment to service and will return more frequently if an operation provides a pleasurable dining experience.

DINING ROOM SERVICE FROM THE CUSTOMERS' POINT OF VIEW

Normally, the most frequent complaint of foodservice customers concerning their dining experience is the lack of friendly, competent service. A meal that is expertly prepared can be ruined for the customer if it is not served properly. In a survey of its card holders in 1990, American Express discovered that a shocking 60% complained about the service they received when dining out. Customers are generally frustrated with the lack of service in dining establishments. They are tired of servers who are less than friendly, ignore them for long periods of time, refuse to honor their requests for special orders, and still expect their 15% tip at the conclusion of the meal. Customers want friendly servers who are knowledgeable about the menu without being overbearing, who are willing to provide reasonable variations in the menu, and who bring things to the table before, rather than after, they are needed.

THE IMPORTANCE OF SERVICE

The service provided by the server and the dining room staff is important for two reasons: it determines value for the guest, and it has a direct impact on the dining experience of the guest. The service staff can have a dramatic effect, either positive or negative, on the customers' enjoyment and on whether they will choose to return. It is hard to understand why such an important job is left to a person who earns less than the federal minimum wage and who is normally minimally trained.

DETERMINING VALUE

Generally, the greater the level of service provided to a guest, the more the guest is willing to pay for a meal. Guests appreciate the increased attention and expect to pay more for it. For example, compare a hamburger and french fry lunch at two types of restaurants—fast food and moderate table service. A customer can usually expect the quality of the meal to be approximately the same at both places. The main differences between the two are the level of service provided and the cost to the guest. At the fast-food/quick-service restaurant, the customers have to walk up to the counter to order the food, carry the food to the table, find their own table, and, finally, eat a meal wrapped in paper while sitting in an uncomfortable booth. The moderate table-service restaurant offers much more. It has someone to show the guest to a table, a server to take the order and deliver the food, and generally much more comfortable seats and tables. When deciding where to dine, customers are aware of the amenities each type of operation offers and approximately what they will be charged for those amenities, and they make their choices accordingly. For many people, the value that good service adds is a key point in making this decision.

ENSURING A POSITIVE DINING EXPERIENCE

The job of the attentive, well-organized server is to make sure that the customer has an enjoyable experience while dining. The ideal server is available when the guest needs him or her, but out of the way when not needed. The competent server is available to make a suggestion if needed, but is not overbearing or aggressive. The prepared server knows the menu and is aware of what can be done to accommodate a guest's special needs if called upon. The server spends more time with guests than all other employees of the operation combined and has the greatest impact on guests' favorable dining experiences.

WHAT CUSTOMERS EXPECT FROM SERVERS

Foodservice customers judge their servers in a number of ways, both consciously and unconsciously. Customers expect and deserve more than just an "order taker"; they expect the server to do more than just take the order, deliver it to the kitchen, and transport the food to them. The key points in maximizing customers' enjoyment include the following:

- Accommodation
- Attitude
- Attentiveness
- Timeliness
- Anticipation
- Suggestive selling

ACCOMMODATION

In planning a menu, it is impossible to anticipate every customer's favorite combination. Sometimes guests grow bored with the menu offerings and would like a different combination. If the objective of the operation is to satisfy the guest, certain policies should be established, supported, and communicated to all involved. The server and the kitchen staff should do whatever they can to accommodate guests' desires. Simple things that are not difficult to accomplish, such as switching the sauce from one dish to another, making a grilled cheese sandwich (that is not on the menu) for a child, or cutting portions in half, serve to take special care of guests and show them that the operation cares about them. Most guests appreciate the small accommodations that servers and others make for them. Servers must be provided with a list of items on the menu that can be switched to better accommodate guests.

ATTITUDE

The attitude of a server has a dramatic effect on a customer's dining enjoyment. A server who is cold or indifferent can ruin even the best-prepared meal, whereas an attentive and caring server can often salvage a less-than-perfect meal.

Good attitude is difficult to teach. It is a trait that is, or is not, inherent in a person's character. Good attitude and enjoyment in working in a service position are key characteristics to look for in the hiring process. Although most people have good days and bad days that can affect their attitudes, attitude in the workplace is generally reflective of a person's feeling toward the job. Management must recognize a server's change in attitude toward the job before it interferes with the service provided to customers.

ATTENTIVENESS

Guests like to have their needs attended to while dining out. In fact, it is the reason that many people like to dine out—they prefer to treat themselves by having someone one else take care of them. One of the complexities of the server's role is to be attentive to the guests' needs. An attentive server will acknowledge guests as soon as they are seated at his or her station, even if the server does not yet have time to take an order. The customers are then at least sure that the server knows they are there. Most dining customers become upset if there seems to be a delay with the meal and their server is nowhere to be found. Servers should notify customers if there will be a delay. They should also return to the table shortly after delivering the meal to see that the guests are satisfied and whether the guests need anything else to complement their meal.

A server should also be attentive to any nonverbal signals that a guest may be providing. Most guests are reluctant to complain; an attentive server may spot some indication that a guest is dissatisfied and take steps to solve the problem. A server who notices that a guest has barely touched his or her meal or drink can inquire immediately about the reason and offer to remedy the situation.

TIMELINESS

As mentioned earlier, guests like to be taken care of in a timely fashion. They expect to be greeted at the table, presented with a menu, have their order taken, their courses delivered, and their check brought to the table in a timely fashion. If there is going to be a delay with any part of the service sequence, the problem should be communicated to the guests as soon as possible. Timeliness does not mean that guests should be served all of their courses in rapid succession. For most guests, the serving of courses too swiftly, one after another, is as annoying as long waits between courses.

ANTICIPATION

Servers who anticipate their guests' needs are more effective and can serve their guests better. For the server, the key is to anticipate rather than react; to be alert to guests' needs and potential problems, rather than having to react and being forced to scramble to remedy a problem. The server or host who notices a guest with an infant should assume that a high chair is needed and bring it to the table before the guest has to ask for it. For the guest who orders a dish served with french fries, the server should bring the catsup to the table after the dish is ordered but before the food is delivered, rather than forcing the guest to ask for it while the fries become cold.

SUGGESTIVE SELLING

Some guests arrive at a foodservice operation with their minds already made up as to what they are going to order, some make a decision after reviewing the menu, and some cannot quite decide and need some assistance. The server who is knowledgeable about the menu should have no problem suggesting menu items to guests once he or she listens to some of their likes and dislikes. Both server and guest benefit; the server receives a larger tip for an increased sale, and the guest enjoys a meal that he or she probably would not have ordered without the suggestion. Suggestive selling also encourages guests to order accompaniments they otherwise may skip, such as wine or perhaps an after-dinner cordial or liqueur.

THE ROLE OF SERVERS

The server is responsible for a number of tasks. The better the server is at performing these tasks, the better at serving the guest he or she is, and the more efficiently the foodservice operation works. The tasks include guest-contact duties designed to ensure a pleasurable dining experience and that help provide efficient delivery of menu items, thus indicates the time it took for the cooks to prepare the food versus the time it took for the server to pick up the food.

TEAMWORK

When not busy taking care of their own guests, servers should be available to aid other employees. A very frustrating response to a guest's question is "Sorry, this isn't my station. I'll find your server and tell him that you need something." Servers should be instructed by management to treat all guests dining in the restaurant as they would guests dining in their own homes. Management should unify all employees in the common goal of taking care of the guest, regardless of whose station the guest is seated in. One restaurant chain has a system called the Flying Food Show. Whenever an order of food is completed, the first server who arrives in the kitchen is responsible for taking it to the table, regardless of station. This system keeps servers from waiting in the kitchen for their food orders to be completed, gets the food to the customers more quickly and at its peak quality and freshness, and encourages a spirit of camaraderie and teamwork within the operation—all parties benefit.

FINISHING FOOD PREPARATION

The role of the server in some operations includes plating and possibly dressing salads, pouring soft drinks, plating and garnishing desserts, dishing up soup, and garnishing entrée items. The problem with having servers complete such tasks is that it takes them away from their primary role—taking care of guests.

MENU KNOWLEDGE

To do their jobs properly, servers must know the menu. The basics, such as how a dish is prepared (baked, broiled, sautèed, homemade, fresh, or frozen), how it can be altered, if at all, and which items are low-fat items, should be included in the server's training manual, and servers' knowledge should be tested on a regular basis. Daily tastings of the menu items, specials of the day, and wines from the wine list, can allow servers to better describe and suggest menu items, as well as to detect a problem with an item before it is served to a guest. The server must know the specials, the soup, and the vegetable of the day, prior to going on the floor to work, in order to answer guest's questions without having to check in the kitchen.

MERCHANDISING

A menu serves as a "silent salesperson" for an operation, providing the guest with its daily offerings. The server has an opportunity for a much more effective and personal level of selling. With the proper approach a server can significantly influence guests' purchases. **Merchandising** skills are easily taught, and their effectiveness is easily measured. Many operations offer incentives, such as meals and cash, to encourage servers to suggest high-profit appetizer and dessert items.

TYPES OF SERVICE

There are two main categories of service—**self-service** and **seated service.** Within each category are a number of variations. Seated service (food is brought to the seated guest), generally considered the more formal of the two, has a number of variations, including American, English or family style, and French. The two main types of self-service are buffet-style and cafeteria-style.

SEATED SERVICE

Customers generally prefer to have their food brought to them rather than having to get it themselves. Some customers feel that the luxury of having food served to them is one of the reasons to dine out. The advantage of seated service is that it is more formal and elegant; the disadvantages are that it requires more labor and generally results in a higher cost to the guest, takes more time for large numbers of people to dine, and generally offers a more limited number of menu choices.

American Service

The seated service most commonly used in this country is **American service.** The food is prepared and plated in the kitchen, then brought to the dining room and served by the server. The standard for American service is to serve solids from the left with the left hand, and beverages from the right with the right hand. Used dishes and tableware are removed from the right. The advantages are that this type of service is efficient, fast, and requires a minimum of equipment. Servers are able to serve many guests with minimal training. The disadvantage is that it is not very formal or elegant.

The procedures for setting a table for American service are as follows (see also Figure 10.1):

1. Place **silencer cloth** on the bare table. This provides a slight cushion for the dining guest and allows the dining room staff to change the table during service without exposing the bare table.

2. Place a clean tablecloth on the table, making sure it is centered and not allowing any part to extend down too far where it might get in the way of the guests. Some restaurants place a top cloth on the tablecloth to make changing easier during service.

3. Place the sugar bowl and the salt and pepper shakers.

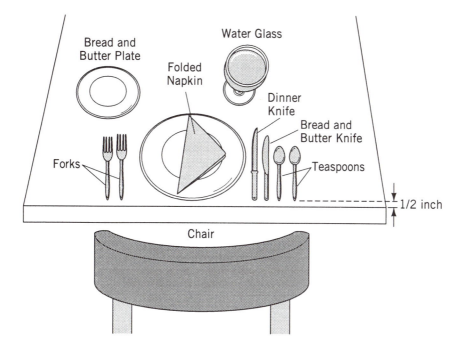

FIGURE 10.1 American service place setting.

4. Place the tableware for each guest:

- A folded napkin in the center of the place setting in front of the guest, 1/2 inch from the table edge;

- Two dinner forks to the left of the napkin (tableware should be placed 1/2 inch from the table edge);

- Dinner knife, butter knife (with the cutting edges facing toward the plate), and two teaspoons to the right of the napkin;

- A bread and butter plate at least 1 inch above the top of the fork—there is an option of placing the butter knife across the bread and butter plate; and

- A water glass, slightly to the right of the tip of the dinner knife.

The tableware is placed on the table as part of the setup, and prior to the guests' arrival at the table. The tableware used for a course is removed after the course is completed or, if it is determined to be an extra place setting, once the guests have been seated.

BASIC RULES OF AMERICAN SERVICE

- All solid food is served from the guest's left with the left hand.
- All beverages are served from the guest's right with the right hand.
- Dishes are cleared from the guest's right.

Adherence to this system provides consistency and allows the server to avoid bumping the guest. These rules are designed for guests dining at a table and require modification for guests dining in a booth.

When seating guests, the server should:

- Seat guests as soon as they enter the dining room, or at least acknowledge them, and then seat them as soon as possible.
- Greet the guests when he or she arrives at the table, welcome them to the restaurant, introduce him- or herself, present them with the menu, tell the guests about any specials for the day or any dishes on the menu that are not available, and ask them if they would like anything to drink or anything from the bar (if available at the operation).
- Fill their water glasses and remove extra place settings.
- Leave the table to pick up the customers' cocktails and to give the guests an opportunity to look over the menu.

When taking the order, the server should:

- Deliver the cocktails (if any have been ordered) from each guest's right.
- Take the guests' order. Most operations have a standardized system for numbering the customers at the table based on a geographical portion of the dining room, such as the front door or fireplace, so that any employee can deliver food to the proper guest.
- After the guests have ordered, offer accompaniments, such as appetizers or wine, to increase the guests' dining enjoyment and, it is hoped, the server's gratuity.

- Remember that the goal of the operation is to satisfy the guests. Try to accommodate guests' needs and desires. Be aware of the policies of the operation so that "special orders" can be offered or accepted.

- Serve the bread and butter. A general policy is to deliver the bread and butter after the order is taken to prevent guests from filling up on bread instead of ordering appetizers.

- Deliver the order to the kitchen. Double-check with the kitchen that all items ordered are available. While the food is being prepared, anticipate the food items and tableware the guests will need for their meal and take them to the table before the food is served. Items such as catsup for french fries, mustard for a hamburger, and a steak knife for steak are examples of things that guests are likely to need.

When delivering the food, the server should:

- Deliver the appetizer, soup, or salad. This course serves two purposes—it stimulates the guests' appetites, and it occupies their time while the main course is being prepared. Soups and salads are sometimes made available for the server to plate up, thus freeing the kitchen staff. Depending on the procedures of the operation, the kitchen may be notified when the soup or salad is delivered. The kitchen staff then knows how to time the preparation of the main course so that it is ready when the guests finish the first course.

- Clear the first-course dishes and tableware, from the right, to make room on the table for the main course.

- When in the kitchen to pick up the main course, collect all of the items needed to serve it—butter, sour cream for a baked potato, serving utensils—prior to picking up the food. The main course is plated by the kitchen, but the server may need to garnish the plates before delivering them. The server makes the last check of the food prior to the guests' receiving it. It is important to see that the correct order is served (not another server's), that the order is complete, and that the quality of the food meets the standards of the operation.

- The food is usually carried to the dining room on a tray, so the server needs to pick up a tray stand to set the food on while serving the main courses to the guests.

- Serve the food to the guests using the numbering system noted when taking the order. Women are usually served first, then men. The main course is served from the left with the left hand.

- Replenish water glasses and ask the guests whether they would like a refill of their other drinks.

- Return to the table within 3 to 5 minutes after delivering the food to ensure that everything is satisfactory and that the guests have everything they need to enjoy their meal.

Toward the end of the main course, the server should:

- Keep an eye on the guests while they are eating to see if they need anything else. When they have finished eating, remove the main course dishes from the right with the right hand.

- Remove items from the table that are not needed for dessert, such as salt and pepper, extra silverware, etc.

- Use a folded napkin to remove crumbs and food scraps from the table.

- Ask the guests if they would like a refill of their beverage, coffee or tea, or an after-dinner drink.

- Suggest some desserts, offer the dessert menu, or present the dessert cart or tray (if used). The visual presentation of desserts is usually very effective.

English Service

English service (host service or **family service**) is less common in commercial foodservice operations. The food is prepared in the kitchen and taken to the dining room in serving platters and bowls by the dining room staff. The food is then plated in the dining room by either the host of the table or by the guests in general. The advantages of this type of service are that it is quicker and easier for the kitchen staff, and each guest at the table is allowed to choose the portion he or she wants. Another advantage for some groups of dining customers is that the meals are generally served as "all you care to eat" for one price. The disadvantages are that everyone at the table is limited to the same menu choice, and that this type of service is not very elegant.

Servers and the kitchen staff must work together to ensure that the proper amount of food is brought to the table. A problem with English service is that when food is returned from a guest's table it cannot be delivered to another table, for obvious reasons of sanitation. Servers may have a tendency to bring large amounts of food to each table in order to limit the number of return trips. This practice, although limiting the number of trips by the server, can lead to great waste of food. The server should limit the amount of food brought to the table, explaining to the guests that he or she will be more than glad to bring as much food as they want, and that the food is remaining hot and fresh in the kitchen.

French Service

It is not surprising that the most graceful and formal style of service originated in Europe. Europe's ruling classes have always been known for formality and elegance, whereas America is better known for its efficiency. **French service** is characterized by the use of silver serving pieces, with the heating or final cooking and garnishing taking place, within view of the guests, in the dining room on a side table *(guerdon),* and the food plated by the server.

French service is by far the most elegant of the seated services, although its use is on the decline on both sides of the Atlantic owing to its need for highly trained service staff and the great expense of the equipment needed. The emphasis is on taking care of the guests' needs. Service professionals are much more highly regarded in Europe than in this country. Waiters, who are more likely to be male than female, attend one of the many waiter-training schools to learn the necessary skills. Upon successful completion of the course and with a few years of experience, the student becomes a *commis de rang.* A *commis de rang* must then serve an apprenticeship under a *chef de rang* (professional waiter) in order to move up to the headwaiter position.

Dining room stations are usually waited on by a team of two servers: a *chef de rang,* or head waiter, and a *commis de rang,* or assistant waiter. They work as a team to take care of the guests' needs. The *chef de rang* always remains in the dining room, whereas the role of the *commis de rang* is to go to the kitchen to pick up the food or to return soiled items. French service requires the most highly trained staff of the three seated services because, in addition to serving the food, the *chef de rang* also finishes the preparation of the food items at the guests' tables. This requires two pieces of special equipment not needed in other types of services: *a guerdon,* or cart, to transport the food and a *rechaud,* or small stove, to cook the food in the dining room. The food is prepped and, depending on the item, partially cooked in the kitchen. The *chef de rang* adds the final touches to the items in the dining room and may finish with flaming the items with brandy to catch everyone's attention.

The advantages of French service are that it is very stylish and formal and that the guests like to see their food prepared before them and consider it a form of entertainment. The disadvantages are that this type of service is expensive in labor and equipment, requires a highly skilled staff in the dining room, and necessitates wider aisles, which means fewer tables in the dining room.

The duties of waiters in French service are included in the following list.

Chef de Rang

- Seat the guests if there is no host for the dining room.
- Take the guests' orders.
- Serve the beverages.
- Finish the preparation of food at the table in view of the guests.
- Plate and garnish the food on a heated plate.
- Present the check to the guests and collect the payment.

Commis de Rang

- Receive the order from the *chef de rang* and deliver it to the kitchen.
- Pick up the food in the kitchen on a silver tray and place it on the *guerdon,* or side table.
- Serve the plated food to the guests.
- Assist the *chef de rang* as needed.

SELF-SERVICE

There two categories of self-service: **buffet service** and **cafeteria service**. Self-service, whereby customers pick up their own food, is considered much less formal than seated service; nevertheless, it offers benefits to both the foodservice operation and the guests. Self-service provides the opportunity for more guests to be served in a shorter period of time and generally offers a greater selection of menu items than does seated service. The foodservice operation benefits because of its ability to serve a greater number of people faster and its need for fewer people to work in the dining room.

Buffet Service

Buffet service is used in a number of different settings—banquets, brunches, salad and dessert bars—and as a restaurant concept. Buffets are used frequently by hotels for holiday meals, as a quick alternative for breakfast, and when they need to serve a large number of people in a hurry. Catering operations use buffets for banquets so that they can serve guests faster and offer several meal choices. Imagine how difficult it would be to select one menu item that would appeal to all of the attendees of a large social function. Some foodservice chains offer salad, pasta, taco, and dessert bars, either as their entire menu or to supplement an à la carte menu in an attempt to offer a greater variety of choices to their guests.

The major advantage of buffet service is that it gives customers greater choice and variety in their meals and portions. The great popularity of the salad bar in the 1970s was due primarily to the fact that for the first time customers, rather than the foodservice operator, chose exactly what and how much was in their salads. Buffets offer the same opportunity—customers, rather than the menu planner, choose what they would or would not like to eat.

The disadvantages of buffet service are that a significant group of people prefer to be served rather than having to serve themselves, and that most food deteriorates in quality immediately after it is prepared. There are also a few potential sanitation problems. Because the food remains out, in open view and within reach of the guests who are serving themselves, there is a great possibility for contamination and cross-contamination. Customers can touch food with unwashed hands, cough, or sneeze, thereby spreading germs, even though there may be sneeze guards in place; or they may use the same serving utensils for two or more items, thereby transferring bacteria from one dish to another. Flies and other disease-carrying insects can land on the food, and there is also the possibility for the food to sit in the TDZ (temperature danger zone) too long, again risking bacterial problems. A major consideration for the foodservice operation is portioning. In seated service the portioning of the food is done to rather exact standards in the kitchen, which allows for fairly accurate costing. Buffets permit guests to choose their own portion sizes and menu items, making it very difficult to determine accurately the cost of their meals. Another problem is that, in theory, there is nothing to stop guests from loading their plates with the higher-priced items in the buffet. The layout of the buffet is usually such that the more expensive items are placed at the end of the line and near the back to make them difficult, but not impossible, to reach.

Cafeteria Service

Cafeteria-style service is used primarily in noncommercial food-service—schools, hospitals, penal institutions, and other places where large numbers of people have to be served in a short period of time with a minimum of labor. Cafetaria service is also used in some commercial restaurant chains, primarily in the southern regions of the country. The setup of the food lines has evolved recently from a **straight-line system** to a **scramble system** (see Figure 10.2).

Straight-Line System.   The original setup of service for cafetaria-style facilities was a single straight line. The patron would start at the beginning, pass all items that were being served, and finish at the end. The advantages of this system are that it requires less staff and takes up less floor space. The disadvantage is that the line tends to back up because all patrons have to pass all of the offerings, regardless of whether they want them or not.

Scramble System.   Cafeteria-system designers decided that if they "scrambled" the various food stations—hot entrées, deli, beverages, desserts—in the food area rather than placing them in a straight line, more customers could be served more quickly and the lines would be shortened. The theory is that if a patron sees a long line at the entrée station, he or she can bypass that station for the time being and go to the beverage station where the line is not as long. The disadvantages of the scramble system are that it takes up more floor space and it requires more personnel to staff.

An adaptation of the scramble system that has become popular in commercial foodservice settings, especially in shopping malls, is the food court. **Food courts** are set up with an assortment of fast-food restaurants on the perimeter and tables in a common center. The advantage of this type of foodservice for a shopping mall is that it allows a great variety of food choices for the dining guests. It also allows a group of friends who enjoy dining together, but have

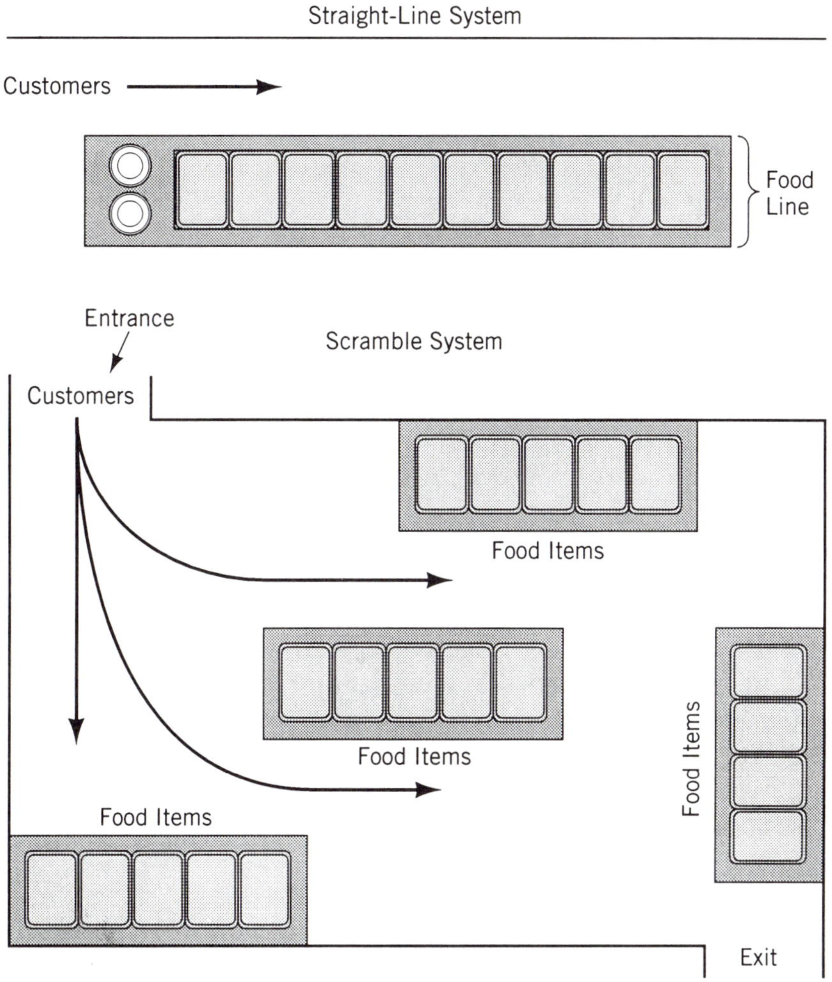

Straight-Line System

Customers ⟶

Scramble System

FIGURE 10.2 The straight-line and scramble systems.

trouble deciding where to dine, to sit together while eating food from a variety of restaurants.

THE OBSTACLES TO GOOD SERVICE

The problem of poor service in American restaurants is a complex one. Customers want and even demand better service; restaurateurs understand the importance of providing good service to their guests. The question then is, what are the obstacles facing restaurants in their pursuit of providing good service to their guests?

The position of server in Europe is held mostly by men and is highly respected. The position of server in this country, for the most part, is looked down on, and is filled by people who are waiting tables while waiting for something else—college students, fledgling actors, and so on. The lack of dedication to a career of waiting tables can have an effect on the quality of service.

To further complicate the matter, American employers are allowed to pay restaurant servers a wage below the federal minimum. The employee who spends the most time with the guest and who has the most dramatic effect on the guest's dining enjoyment is rewarded by a below-minimum wage. Such an employee does receive tips, but management cannot expect to attract people who are serious about service if they are offered such a low wage.

Training is very important to a server's ability to provide good service to guests. Today's poor service has been blamed on the method of training, or lack of it, that is given in most foodservice operations. Managers cannot expect servers to be competent and efficient at their jobs if their training is limited to the location of their station and how to fill out a ticket for food. Servers are frequently trained by **shadowing** or **trailing** an experienced server or server-trainer. The new server is given an opportunity to pick up the good habits and methods of the trainer; the problem is that he or she also has an opportunity to pick up the bad habits. The success of shadowing is limited to the learning the mechanics of serving—how to fill out a ticket, how to set a table properly, and how to do the side work. The key components of the server's job are not usually addressed, such as suggestive selling, anticipation, and accommodating the guests' needs.

SUGGESTIONS FOR IMPROVING AND MAINTAINING SERVICE

Most important in building an effective program of quality service is the hiring process. Management must be careful to hire people who possess the characteristics that will help them to become good servers. When hiring, a manager should look for people who have a good attitude and an understanding of the components of service. Training someone in the mechanics of service is much easier than training or influencing a persons' attitude.

Servers need a formal training program that covers both the mechanics and the customer-service components of the job. Instructional material must be presented to and shared with the server regarding the items included on both the food and wine menus. This is important in helping the server to be knowledgeable about the menu. The server should also be instructed as to allowable alterations of menu items, how to inform guests about particular items, and how to make suggestions for accompaniments.

To further reinforce the server's knowledge of the menu, restaurants should hold pre-meal meetings to discuss the meals and to provide samplings of the specials and regular menu items. The sampling of menu items allows the kitchen to discover problems before the food is served to guests and enriches the servers' knowledge of the menu. Customers frequently ask a server's opinion of a food item and may be quite annoyed with the response, "All of the food is good here." Customers already know that, or they would not be dining at this facility. They are requesting an opinion that can be given only by someone who has eaten the particular food item. Wines on the wine list should also be tasted and matched with the items on the menu they will complement, again to allow the server to recommend the wines more knowledgeably.

Gary Penn, owner of the Professional Waiters School in Los Angeles, has some definite opinions on the state of service in America and how to improve it. He believes that such problems can be traced to the foodservice managers who do not spend enough time training servers properly. He offers the following suggestions for restaurateurs who want to increase the level of service in their dining rooms:

- Role-play all service situations, especially any problems the wait staff may have encountered or may see as potential problems.

- Develop a training manual that each waiter can take home and read.

- Develop a food and wine manual based on the menu and wine list. ("You can't assume every waiter knows what a Bernaise sauce is—or even knows how mayonnaise is made," says Penn.)

- Put greater emphasis on hiring. Look more closely at an applicant's personality, energy, and ability to handle pressure. Consider less where the applicant has worked before, and whether he or she is good-looking.

SUMMARY

The service provided with a meal has a dramatic effect on a customer's enjoyment of the dining experience. For a server to have greater influence on the customer's enjoyment, that server must be more than simply an order taker and food deliverer. The role of the server should include anticipating guests' needs before they have to ask for something, accommodating their needs, and communicating with other servers and staff.

The negative perception of the profession of waiting tables in this country is one of the obstacles to good service in American foodservice operations. Other obstacles include the below-minimum wage servers are paid and the poor training programs of most operations. The foodservice industry in America needs a revolution in the service area similar to the revolution and excitement that have occurred in the area of food and dining out in general.

Among the solutions to the problem is to have foodservice operations spend more time training their servers in the mechanics of serving, the key aspects of their role, and knowledge of the products they will be selling and serving.

KEY TERMS

accommodation	family service
attitude	French service
attentiveness	*commis de rang*
timeliness	*chef de rang*
anticipation	*guerdon*
suggestive selling	*rechaud*
teamwork	buffet service
merchandising	cafeteria service
self-service	straight-line system
seated service	scramble system
American service	food courts
silencer cloth	shadowing
English service	trailing

DISCUSSION AND REVIEW QUESTIONS

1. What are the two main reasons that the service provided by the dining room staff is important to the success of a foodservice operation?

2. How does the level of service provided determine value for the guest? Give an example.

3. Name five of the key points of a server's job, in addition to transporting food to the dining room. Give an example of each, and explain each example.

4. Fill in the following table:

	Type of Service		
	French	American	English
Where is the food cooked?			
Where is the food plated?			
From which side of the guest is the food served?			
From which side of the guest are dishes cleared?			
Advantages			
Disadvantages			
Type of foodservice operation where this service would be found			

5. Which type of service uses the *chef de rang* and the *commis de rang?* Name two of the responsibilities of each. What are the advantages and disadvantages of a two-server system?

SUGGESTED READING

Casado, Matt A. *Food and Beverage Service Manual* (New York: John Wiley & Sons, 1995).

CHAPTER ELEVEN

SAFETY AND CLEANING

CHAPTER OBJECTIVES

Upon completion of *the "Safety Management" portion of this chapter you should be able to:*

- Recognize the potential dangers of working in a foodservice establishment.
- Understand the causes of most foodservice accidents.
- Explain ~~three~~ reasons for a safety management program.
- Set up a basic safety management program.
- *State* ~~Compare~~ the manager's and employees' responsibilities in a safety management program.

Upon completion of *the "Cleaning Program" portion of this chapter you should be able to:*

- Explain the importance to customers of cleanliness in a foodservice establishment.
- State two reasons that self-inspection of an operation is important.
- Name the four main components of a master cleaning schedule.
- Understand the basics of selecting the cleaning tools and equipment necessary to implement a cleaning program.
- State the keys to the successful implementation of a cleaning program.
- Understand the importance of follow-up in maintaining the program.
- Explain the causes of failure in a cleaning program.

The first part of this chapter focuses on the causes of accidents and how to develop a safety management program. The second part discusses organizing and implementing a cleaning program appropriate to a particular type of foodservice operation.

SAFETY MANAGEMENT FROM THE CUSTOMER'S POINT OF VIEW

Customers assume that when they dine in a foodservice establishment, the operation will be free from physical hazards. Most of the effort of a safety management program is out of view of the guests. The principles and guidelines established for the program are primarily designed to protect the workers of the operation, although an operation that is made safe for the employees is also safe for guests. Employees trained in safe work techniques throughout the operation will be able to use those techniques in the dining room as they serve with guests.

Foodservice operations, by their very nature, are dangerous places in which to work. The sharp knives, grease, steam, hot surfaces, and the general rushed pace provides an environment conducive to accidents. By definition, accidents are random occurrences that cannot be avoided. A premise of this chapter is that what are called *accidents* in foodservice operations are not really accidents according to this definition. Most foodservice operation mishaps, which are commonly called accidents, can be avoided and prevented through an organized safety management program. Properly training employees to be aware of the potential dangers of the equipment they operate and the environment in which they work can reduce the chances of accidents occurring.

This part of the chapter examines the causes of accidents and then discusses how, by forming a safety management program, the number of mishaps in the workplace can be reduced.

THE REASONS FOR A SAFETY MANAGEMENT PROGRAM

In addition to the moral obligation of an operation to provide a safe workplace for its employees and customers, business owners and operators must consider legal, monetary, and staffing issues when planning a safety program.

LEGAL ISSUES

Business owners are obligated by law to provide a safe environment for their employees. The Occupational Safety and Health Administration (OSHA), is the federal regulatory agency in charge of enforcing the standards related to the safety and health of U.S. workers. OSHA has the power to fine employers for unsafe workplaces.

MONETARY ISSUES

Insurance premiums are often based on the safety record of an operation. The more accidents an operation's employees have, the more insurance claims they have to file, the more the insurance company has to pay, and the higher the insurance cost. Several companies have found that instituting a safety management program is a beneficial way to reduce costs. Employees and guests can sue an operator if they can prove that the operator was negligent and did not take sufficient care to provide a safe place to work or eat. Business owners who knowingly allow unsafe conditions to exist in their operations can be liable for accidents they cause.

STAFFING ISSUES

One of the leading problems that foodservice operations face is the labor shortage. Unsafe working environments and accidents reduce the number of possible employees and decrease the morale of the staff. A serious accident that occurs in a foodservice operation, especially if it is caused by a hazard known to management, can have detrimental effects on the spirit of the workers. A severe reduction in morale can lead to reduced production and increased turnover.

Finding sufficient staff and enough time to train those hired is a major concern of most operators. An investment in a safety management program can not only reduce costs, but can also reduce the time and effort needed to hire and train new staff to replace those lost to injury.

COMMON INJURIES

Cuts, slips and falls, and burns are common injuries among foodservice workers. Kitchens are dangerous places. Knives in people's hands and attached to machines; heat-generating equipment necessary to cook food, such as ovens, stoves, broilers, and steam-generating equipment; and cooking grease are all potential injury-causing components of most kitchens. These hazards are further compounded by the rapid pace of activity in most kitchens. Foodservice operations often seem to be behind schedule or short-handed, so that workers are forced to hustle at a hectic rate to complete everything prior to meal service.

WHAT CAUSES ACCIDENTS?

The term *accident* implies a random act or occurrence, or something that is unavoidable. Most such occurrences in foodservice operations are preventable. The prevailing theory is that with the understanding of the causes of accidents, managers can be more effective in developing ways to prevent them.

Lack of Training

The rapid turnover of foodservice workers in many operations often means that employees do not receive the adequate training needed to work safely in these potentially dangerous places. Workers who are not properly trained in equipment operation can cause damage both to the equipment and to themselves. Those who are properly trained in the operation of the equipment they will be using on the job will both increase productivity and reduce the possibility of injury and damage to the equipment.

Fast Pace of the Kitchen

Kitchens and foodservice establishments often operate under time pressures. Managers and other staff members know that providing guests' meals in a rapid and timely manner is important to the success of their operations. Thus, the pressure of deadlines, coupled with the general shortage of help, forces workers to do their jobs more quickly than may be safe. To help reduce accidents, managers should discuss with employees the dangers in working too fast.

Potentially Dangerous Equipment

The nature of foodservice equipment is potentially dangerous. Preparation equipment is designed to cut, chop, or slice food items, and workers can be injured if such equipment is not used properly. Equipment that produces heat to cook food can easily burn an unsuspecting or ill-equipped cook. Most foodservice equipment, although potentially hazardous, is designed with a number of safety features. Unfortunately, workers sometimes bypass these safe guards because of time pressures or because of improper training. Workers should be educated about the potential dangers of the equipment they work with to increase their awareness and help to prevent injuries.

KEYS TO A SUCCESSFUL SAFETY MANAGEMENT PROGRAM

Most accidents in foodservice operations can be prevented. A coordinated program supported by all employees and management is essential. Operations must be proactive, rather than reactive. In other words, managers must anticipate problems and try to prevent them before they occur, rather than wait for mistakes to happen and then react.

TEAM APPROACH

For a safety program to be successful, it must be supported by all levels of management and all employees. The entire operation must participate and be involved. Nothing boosts morale and demonstrates the support of upper management for a safety program more than a general manager of the operation, upon seeing a server spilling a drink on the way to the dining room, picking up a mop him- or herself, rather than calling a bus person to do the job.

INCREASED AWARENESS

Studies have shown that awareness of potential problems and accident areas helps to reduce incidents. Operations should have monthly safety meetings to promote awareness, to review accidents, and to determine ways to prevent them. Safety issues should be incorporated into employee meetings to help maintain awareness.

CONTESTS TO ASSIST IN MAINTAINING AWARENESS

Sharing with employees some of the savings derived from a safety program is an effective way to maintain awareness. Dividing the operation into teams, and then running contests, tends to turn all employees into "safety police." Awarding money and prizes to workers who work safely is a great motivator in sustaining the momentum of the safety program. The cost of the rewards can easily be recouped through the savings realized by the reduction in accidents.

THOROUGH EMPLOYEE TRAINING

Management cannot rely on employees' prior knowledge and experience regarding safety practices. Foodservice equipment is too potentially dangerous and the turnover is too high in such operations. Employees must be trained and checked regularly on the use of the equipment in order to minimize problems. Operators should also have formal, written safety and operational procedures for all pieces of equipment in the establishment. These written procedures will assist in training and ensure that all employees are consistent in their safe use of the equipment.

EVALUATION OF EMPLOYEES' SAFETY AWARENESS

Workers should be evaluated on their safety awareness and participation in safety programs. Employees who exhibit increased safety awareness should be recognized and rewarded, and those who exhibit exemplary safe-work skills should be commended to the rest of the work staff.

EMPLOYEES AS SAFETY POLICE

If all workers in an operation are looking for potential safety hazards, the number of accidents will be reduced. A team spirit approach is beneficial to the success of the program. With all employees acting as safety police, rather than making this the responsibility of managers only, safety will be increased dramatically.

RESPONSIBILITIES OF EMPLOYEES AND MANAGEMENT FOR SAFETY

The responsibility for safety management in a foodservice operation must be shared by managers and employees. The program will work only if both groups realize and understand the benefits of the program and are willing to participate in it.

EMPLOYEE RESPONSIBILITIES

Keeping Alert and Paying Attention

The inherent dangers of a kitchen require workers to be aware of what is going on around them. Distractions should be kept to a minimum so that they can concentrate on the tasks at hand and thereby work safely. The operation should be set up in such a way that an employee can spot potential safety problems before they occur.

Exercising Common Sense

Common sense is practically impossible to teach; however, during employee meetings, managers can plan a question-and-answer-type presentation of potential hazardous situations. Employees can be presented with hypothetical situations in which they practice "thinking safely." Through this exercise errors in judgment or pre-conceptions can be corrected before disaster occurs. People need to think when they reach a questionable situation, and they need to be able to use enough sense to avoid an accident. *Remember the point that cannot be stressed enough: Most accidents can be avoided.*

Assisting Other Employees

Workers must help each other to avoid accidents and to discover unsafe conditions. Workers looking out for each other will benefit the entire operation by increasing both safety and morale.

Informing Management of Potential Hazards

Managers cannot be in all places at all times. Designating all employees as safety police increases the number of eyes looking for problems and decreases the number of accidents. It is the workers who are using the equipment; they are better situated to discover safety problems and potential hazards.

MANAGEMENT RESPONSIBILITIES

Providing and Maintaining a Safe Workplace

Employees can work only with the equipment and workplace setup provided to them. Managers control the allocation of funds and determine the amount of money available for repairs and replacement of equipment. For a safety program to be successful, management must be committed to providing and maintaining a safe workplace.

Providing and Maintaining Safe Equipment

Kitchen equipment is potentially hazardous when it is in perfect working condition. Safety hazards can increase substantially if equipment is not properly maintained. Proper maintenance will also help to ensure the smooth operation and longevity of the equipment.

Training Workers in Safe Work Habits

Managers must be committed to the proper training of workers regarding safety and safety management. Employees will be only as safe as they are trained to be. Safe work habits should be a key component of any employee orientation and training program.

Following Through on Employee Suggestions

Nothing will stifle employee feedback more then nonaction by management. For example, if an employee reports an unsafe condition to management and nothing is done to rectify the situation, employees will not bother to point out other problems. Any safety program can fall apart if managers do not react to employee feedback and suggestions.

ORGANIZING A CLEANING PROGRAM FROM THE CUSTOMER'S POINT OF VIEW

Customers equate cleanliness with sanitation. If an operation looks or smells unkempt or dirty, customers will assume that it is unsanitary. The following survey results indicate the views of foodservice customers on the importance of cleanliness in the places where they dine. The survey asked customers to name the attributes they consider very important.

People who work in foodservice understand that cleaning is necessary to maintain a functional and healthy operation. It is necessary to clean and **sanitize** items that come in contact with food or customers before those items are reused. In order to maintain a clean and sanitary foodservice operation, management must insti-

**What Customers Want and
Consider Very Important When Dining Out**

Food Quality	92%
Cleanliness	92%
Friendly Service	66%
Atmosphere	66%

Source: Restaurant Business Research, 1992.

tute a systematic cleaning program. Management must be consistent and fair in setting up the program so that all areas of the operation share evenly in the cleaning duties.

Management is the key to any cleaning program. Managers are responsible for setting up a program, ensuring that the necessary supplies are available, and assigning the various duties to employees. However, management's responsibility does not end there. For a cleaning program to survive, management must continue to support it and to make sure that all the areas of the operation are following the program and abiding by the operation's standards of sanitization.

For a cleaning program to be successful, it must be carefully planned. Before initiating a program, managers must determine what needs to be cleaned and when. Once a list is compiled, a master cleaning schedule is developed. The master cleaning schedule includes what is to be cleaned, who is to clean it, when it should be cleaned, and how it is to be cleaned. The next step is to determine what materials and resources are necessary to perform the jobs. When all this is accomplished, the program must be introduced to the employees in a way that ensures their cooperation. The program will be effective only if the employees support it. Managers must make a special effort to "sell" the program to the staff.

EXAMINATION OF THE OPERATION

The implementation of a cleaning program must begin with an accurate and thorough list of what needs to be cleaned. Most employees do not relish the idea of cleaning and will gladly skip cleaning items that are not on the list. The inspection of the facility and compilation of the list is best accomplished by having more than one person tour the operation to determine the cleaning needs. This is a good time to involve supervisors and to use their input. This will help to gain their support in the implementation stage. The more people involved in the process, the less chance something will be missed. The customer's point of view must also be considered—the customer must enter an operation that is free from visible soil and properly maintained.

Once the lists from the various inspections are compiled, further details should be considered. The successes of the previous cleaning system, if any, should be examined and included in the new program. The failures or shortcomings of the previous system should also be examined to see where it failed and what can be done for improvement. The hours of operation, as well as the patterns of business, will dictate when some of the equipment can be cleaned.

DEVISING THE MASTER SCHEDULE

The information gathered for a cleaning program should be compiled in a **master cleaning schedule** so that is easily understood by the people who will follow it. The program will fail if the workers do not understand what is expected of them. The more specific the list, the greater the chance that it will be followed and the greater the probability that the desired results will be achieved.

For example, imagine the confusion that would *confront a worker* who reads that his or her cleaning assignment for the shift is to "Clean the walk-in," without any further instruction or without having been trained in the procedure. Does *cleaning* mean the inside? The outside? Or the inside and the outside? Does this instruction mean to take the things out of the walk-in, or to clean around them?

Without instructions specifically detailing what needs to be done, time is wasted and no one can be sure of what to expect. Thus, to ensure compliance and reduce confusion, it is important that items on the schedule are broken down into four key components:

1. What is to be cleaned?

2. Whose job is it to clean it?

3. When should it be cleaned?

4. How should the cleaning be completed?

WHAT IS TO BE CLEANED?

A schedule must include the items that need to be cleaned. To increase the schedule's effectiveness, the items that need to be cleaned should be organized according to the areas in which they are located. Within each area, the equipment should be listed in the order in which it is found in the work space, in clockwise fashion, or in the order in which a person would most naturally, or most logically, clean the area. This will make the instructions easier to follow and increase the chance that no item will be skipped.

Self-inspection of the operation is important in putting together the cleaning schedule and in maintaining cleanliness. Management, as well as staff, should inspect the operation on a regular basis to monitor the sanitation level. It is particularly important to inspect the areas that guests see; it is always better for staff members to find something objectionable than to have customers see or smell it. Employee involvement in the process will increase their acceptance of the cleaning program.

WHO IS TO CLEAN IT?

Employees must be held **accountable** for their cleaning responsibilities. The schedule should include specific job positions that will be responsible for cleaning certain items and areas. Suppose a cleaning task were uncompleted or done incorrectly. If assignments were not given to specific people, then determining who is not keeping up with his or her cleaning duties would be difficult.

Employees should be held responsible for cleaning the areas in which they work and the equipment with which they work. This encourages them to adopt a **clean-as-you-go** policy and to keep their areas cleaner. Responsibility for larger items that are used by a number of people should be split evenly among those using the equipment.

One of the keys to the success of a cleaning program is the fair distribution of work among all areas and shifts in the operation. An unfair distribution of responsibilities among shifts can increase the natural dissention between the groups. Employee input on an equitable split is one way to reduce the problem.

WHEN IS IT TO BE CLEANED?

The question of when cleaning tasks should be performed is best determined by the manager and the supervisor of an area. Scheduling within a month and within a shift are both important. Major cleaning projects should be scheduled for regular intervals so that soil does not accumulate and reduce the safety and effectiveness of the equipment. Managers should consider the busy and slack nights of the week, as well as the busy and slack times of the day, in order to schedule cleaning assignments.

It is important to schedule cleaning tasks so that they will not interfere with serving guests or contaminate food. For example, service in some restaurants seems to suffer immediately before closing time when workers are more concerned with cleaning their areas, so that they can go home, than with serving guests; nor is a lull in lunch business a good time to de-lime the steam table, which may also risk contaminating the food stored there.

HOW IS IT TO BE CLEANED?

Employees need a clearly written and illustrated guide on how to perform their cleaning tasks. The guide, which can be in the form of a poster, should be placed close to the area where the cleaning is to be done. This guide serves as both a training device for new workers and a friendly reminder for veteran workers. Conveniently located guides are an easy way to help employees do their assigned jobs properly.

The specific chemicals and tools needed for the job should be included in the guide. The industrial-strength chemicals used in foodservice operations are dangerous, and it should not be left to an employee's discretion as to what chemical should be used when. An untrained, uninstructed employee could think that mixing two chemicals used for cleaning will do a better job then just one, not realizing that the mixture could result in a dangerous or even deadly combination.

The illustrated guides should be made out of a durable and waterproof material to withstand the conditions in a kitchen. The information should be given in Spanish or any other necessary language, as well as in English, so that everyone can understand the material. Some chemical suppliers and companies provide illustrated guides specifically intended for the use of their products in a commercial kitchen. Although these serve primarily as an advertisement for their products, they also prove very helpful for operations that do not have the resources to produce their own.

SELECTING CLEANING TOOLS AND EQUIPMENT

The efficiency and safety of workers can be greatly affected by the materials used for cleaning. Employees will lose interest in performing a cleaning task with equipment or chemicals that do not work or are not available. The lack of safety equipment—gloves, goggles, long-handled brushes—can cause an employee to be burned or hurt while performing an assigned cleaning task; it might also cause him or her to neglect to do that task next time.

The chemicals that are chosen should be checked to make sure that they are suitable for the required job, that they will not cause harm to the equipment, and that they will not leave a residue that may come in contact with food. Manuals for various equipment, such as dishwashing machines, should be consulted to ensure that the proper chemicals are used to maintain their effectiveness without causing mechanical problems. Chemical company representatives often provide valuable information regarding the chemicals needed to maintain a clean and sanitary operation. Such companies can be a helpful resource.

Each piece of equipment used in cleaning should be commercial or industrial grade because of the heavy use it will encounter. The cost is naturally higher, but such materials will be less expensive over time because, if properly cared for, industrial-grade equipment will not have to be replaced as often as lesser-grade equipment. Care must also be taken in the selection of cleaning equipment to ensure that it does not harbor bacteria and recontaminate the surfaces that are being cleaned. Equipment should be labeled so that any used with **toxic** chemicals is not used with food or food-contact surfaces; for example, a plastic bucket used to mix cleaning solutions should not be used to make salad dressing.

THE MANAGER'S RESPONSIBILITIES FOR THE CLEANING PROGRAM

Remember, managers are only as effective as their workers. The resolution of complaints about an unclean operation by a customer or the board of health will ultimately be the manager's responsibility; therefore, the effectiveness and success of the cleaning program is the responsibility of the management team. Managers cannot clean the operation themselves—they must depend on the workers to do that. The effectiveness of the workers' cleaning depends on the forming, implementation, and follow-up of a cleaning program.

To keep a cleaning program operating efficiently, the management staff must support the program. The equipment and supplies necessary to do the job must be carefully maintained in adequate quantity and good repair by the management staff. In addition, management must ensure that the cleaning is being done by checking the clean-up jobs regularly. Workers who know that a manager is too busy to check their work at the end of their shift may leave some tasks undone. When the next shift arrives, those workers may refuse to clean items left by their predecessors, and thus the problem can snowball. The schedule must be flexible enough to be adapted as needed. Managers must continue to inspect and "keep their eyes open" as they tour the operation to spot any problems in the cleaning program and solve them before the system deteriorates.

IMPLEMENTING THE CLEANING PROGRAM

All of the time and hard work spent in devising a master cleaning schedule can be lost if the management staff does not do a good job of **implementation,** "selling" the program to the employees who are going to use it. Time must be taken, and considerable effort must be made, to present the program to employees in a way that enables them to understand and encourages them to participate. Other fatal management errors could be a lack of follow-up and inflexibility in the program and schedule. Management's role is to maintain everyone's enthusiasm for the program and to adapt the schedule as situations change in the operation.

A new cleaning program is destined to fail if management thinks it can simply post the new program and expect workers to automatically follow the schedule. A carefully planned implementation meeting should be held to initiate the program. It is important to explain the purpose of the program so that employees have a better understanding of why each task is being performed and the place of each task in the overall program. Cleaning techniques must be demonstrated to the employees if they are expected to perform the tasks properly.

Follow-up is crucial. Managers must inspect all areas after each shift to ensure that the program is being followed. A grading system may be devised, with awards or prizes distributed to areas or individuals receiving the highest scores. Adherence to the cleaning schedule should also be included as part of the employee evaluation process. Equipment and supplies must be maintained so that employees have the tools necessary to perform their jobs.

SUMMARY

Programs for safety management and cleaning are both essential to ensure a smooth-running foodservice operation. The benefits of each include reduced costs, increased profits, increased employee morale, and reduced turnover. Foodservice managers must be aware of the components of both types of program in order to run a successful operation.

The potential safety problems of a foodservice operation make implementation of a safety management program essential. Employees must be made aware of the likely dangers that are present in the workplace so that those dangers can be avoided. An organized system will make training new employees in safe work practices easier; it will also make the operation a safer place in which to work.

Without specific training, most employees do not clean and maintain foodservice areas to the level required by the board of health. To further complicate the problem, most cleaning is best performed at the end of a shift when employees are in a hurry to leave work. In order to maintain a clean operation, management must develop and maintain a cleaning program. A cleaning program begins with a thorough inspection of the operation to determine what needs to be cleaned. Once the needs are determined, a master cleaning schedule is assembled. This schedule includes who should clean what, when specific areas and equipment should be cleaned, and how each area and piece of equipment is to be cleaned. The schedule provides the employee with all information needed to perform a cleaning task. This is a powerful management tool in any cleaning and maintenance program. The schedule is also a good training tool. A cleaning schedule can, and should, be included in the training program of new employees.

KEY TERMS

sanitize	toxic
master cleaning schedule	implementation
self-inspection	follow-up
accountable	accident
clean-as-you-go	OSHA

DISCUSSION AND REVIEW

1. Name three reasons for establishing a safety management program.

2. Why is the term *accident* inappropriate to describe mishaps in foodservice operations?

3. Name two employee responsibilities and two management responsibilities in a safety management program.

4. How do customers rate the importance of the cleanliness of a place where they dine?

5. Explain the importance of self-inspection in a foodservice operation.

6. Explain the importance of the four main components of a master cleaning schedule.

7. Why is management follow-up crucial to the success of a cleaning program?

8. Describe three factors that can cause the failure of a cleaning program.

SUGGESTED READING

Educational Foundation of the National Restaurant Association. *Applied Foodservice Sanitation,* 4th ed. (Chicago: Educational Foundation of the NRA, 1991).

CHAPTER TWELVE

COST MANAGEMENT

CHAPTER OBJECTIVES

Upon completion of this chapter you will be able to:

- State the difference between fixed costs and variable costs and list examples of each.
- Describe the major categories of costs in a foodservice operation.
- Tell the differences between a *recipe* and a *standardized recipe*.
- State the components of a standardized recipe.
- Describe the importance of accuracy in converting recipes.
- Convert the yield of a recipe to fit a specific need.
- List the advantages of yield tests and quality checks.
- Describe the advantages and disadvantages of two methods for calculating the selling price of menu items.
- State the tangible and intangible factors that management must consider when setting menu prices.
- Describe the ~~advantages of effect~~ive human resources management.
- List the tools that are important in human resource management.

As an essential function for the success of a foodservice operation cost management is second only to taking care of the customer. An operation that does not actively work to control costs is usually headed quickly toward financial ruin. Foodservice operations typically have a low rate of success in their first few years of operation, which is often caused partially by a lack of adequate cost controls in the business. It is difficult for a foodservice business that operates on the rather low profit margin of 5% to 15% to remain in business long if it is selling food at a price too low or if more food is going out the back door with employees than is being paid for by guests. Although it is virtually impossible for them to guarantee success or to eliminate theft totally, controls can be put in place to monitor the operation and indicate the presence of problems. The nature of the foodservice industry makes cost management difficult and contributes to the high percentage of foodservice business failures in the first few years of operation.

There are three main categories of costs in foodservice operations: food, labor, and overhead. This chapter concentrates on the controlling of food costs; labor and overhead cost control are discussed briefly. Food costs, a significant portion of an operation's total costs, are difficult to control because of fluctuations in levels of business and the perishability of food. The keys to controlling food costs are the implementation and use of standardized recipes, along with an accurate determination of the cost of menu items, and the careful calculation of selling prices for menu items. Standardized recipes provide the cooks with accurate measurements of food items and detailed instructions on how to make a dish; they are instrumental in assuring consistent food quality and quantity. Management must have a good idea of the cost of a dish in order to determine at what price to sell it so that the operation can recover those costs. The accurate determination of the cost of some components of menu items can be a problem. The cost of the food is easy to calculate; the problem lies in determining the proportion of labor, utility, and overhead costs that can be accurately allocated to menu items. Once the cost of a menu item is calculated, management must determine what to charge for the item in order to recover the cost of the food and the other costs of running the foodservice operation.

COST MANAGEMENT FROM THE CUSTOMER'S POINT OF VIEW

Cost management, or control, in a foodservice operation goes on mainly behind the scenes and out of view of the customer. The only possible indication to a customer that an operation has failed to institute an effective cost control system is its going out of business. Ineffective cost management often forces companies to continue to raise prices so as to compensate for the loss of money and food. Customers notice and soon resent the spiraling costs and eventually stop patronizing an operation whose prices are too high.

THE MAJOR COSTS OF A FOODSERVICE OPERATION

The major costs of a foodservice operation can be grouped into three major categories: food and beverages, labor, and overhead. Each is essential to the operation of the business, but because their natures each vary, must be examined separately if one is to develop a way to understand and control them.

Within each category, costs can be fixed or variable, depending on their nature. **Fixed costs** such as rent, salaries, and insurance remain the same—fixed—regardless of the amount of business. **Variable costs,** such as food, hourly labor, and utilities fluctuate according to the level of business of an operation. For example, if an operation's business doubles from one month to the next, the fixed costs (rent, salaries) remain constant over the two months. However, because the operation serves twice as many people in the second month, its variable costs (food, part-time help, utilities) will go up according to the increase in business.

FOOD AND BEVERAGES COSTS

The cost of the food and beverages purchased for resale to guests can constitute 30% to 50% of the costs of a food service operation. Food costs vacillate, and the percentages fluctuate depending on the type of operation and the amount of preparation the purchased food items require. The fluctuation of food costs can also depend on additional factors, some of which are controllable by management, and others of which are not.

There are many aspects of the foodservice operation that have an effect on food and beverage costs. Ordering, purchasing, receiving, and storing, if done improperly, can cause food costs to increase. For example, ordering too much food, not negotiating the best possible price with a supplier, not checking the order properly and signing for it—and thereby being charged for an invoiced item that was not received—and storing a product too long without rotating the goods, can cause a reduction in the quality of the goods to a point that the item is unservable. All of these practices affect food cost in one way or another.

In the area of production, several situations can arise that will increase food cost. Food that is cooked too far in advance or over-cooked can be unusable. Any food item that is not served to a guest and paid for has a detrimental effect on the food costs of the operation. A food item whose recipe specifies that it be served in a portion of a particular size that is then served in a larger portion causes increased expense and distorts the food costs of the operation. Accurate portion control is essential for both cost management and consistency of products served to guests.

Serving control, or management of the food items once they reach the dining room, also has an effect on the food costs of the operation. The dining room staff must make sure that all the food that is served is charged to the guests, and that accurate orders are being taken from the guests. Food items that are served to a guest and then returned because of a mistake by the server are unservable to another guest. This costs the operation money that cannot be recovered, thereby contributing to high food costs. Items such as bread, rolls, and butter or margarine that are served to guests must also be controlled. Any uncovered item that is served to a guest but returned to the kitchen because the guest did not eat it must be discarded. Care must be taken with rolls and bread so that guests receive all they want, without leaving any in the basket to be discarded.

Food costs are difficult to control because of several inherent factors that make them unique. Foods are perishable, which decreases the shelf life and the useful life of the products. The variable nature of business in foodservice operations makes it difficult to forecast accurately the number of guests who will eat at the establishment. It is therefore difficult to determine how much food to order. If an operation orders too much food, it may deteriorate in quality, becoming unusable and causing waste that raises the operation's food costs.

LABOR COSTS

Controlling labor costs is a perennial problem for foodservice operators. Labor costs are influenced by several factors unique to the industry.

Foodservice operations are very labor-intensive by nature—the majority of work is done by hand rather than by a machine, thus increasing the need for a large number of workers. Most of the jobs in a foodservice operation require unskilled or low-skilled employees. The high percentage of minimum- or near-minimum-wage jobs makes it difficult to attract a large number of highly motivated, productive, and competent workers. The resulting high turnover rate increases costs to management, which must constantly hire and train replacements. Yet not all foodservice operations are plagued by high turnover rates. Some operators have instituted training and support programs to reduce turnover, saving the operation money and the effort of having to hire staff continually.

Labor costs can also be affected by other factors, such as government-mandated minimum wages. A raise in the national minimum wage raises the labor costs of many foodservice operations that rely on employees who work at this pay scale. The cost of fringe benefits, such as medical insurance and vacations that are provided to employees in addition to their base pay, can significantly increase labor costs.

OVERHEAD COSTS

Overhead costs, or operation costs, include the costs of running the business that are not food or labor expenses. These costs, for such things as rent, utilities, and insurance are. Management must examine overhead costs to determine how it can best economize in these areas.

A LOOK AT FOOD-COST MANAGEMENT

Food constitutes the highest single area of cost, and therefore warrants close scrutiny. The two most crucial components affecting food costs are the use of standardized recipes and determination of the proper selling price. These factors, essential to a foodservice operation, are discussed in detail in the following paragraphs. Kitchen workers must be provided with a written set of instructions and a list of ingredients in order to produce consistent food products. An accurate and systematic method of calculating the cost of menu items is necessary to determine how much to charge for each item in order to cover the costs adequately and thus remain in business.

STANDARDIZED RECIPES

The use of standardized recipes is essential to controlling food costs and maintaining product consistency. The time and effort required to develop a standardized recipe or to adapt one to fit the needs of a particular operation will be well rewarded.

A **recipe** is a list of ingredients and a set of instructions for preparing a food or a drink. Recipes come in many formats and from many different sources. A simple recipe, without detailed instructions or accurate ingredient measurements, may be adequate for people who cook at home where consistency in the quality or quantity of an item may not be crucial. For example, a person may have a friend who makes rice pudding he cannot resist. So he asks his friend for the recipe in order to make this delicious dish at home. The following is the recipe he is given:

Rice Pudding

Sugar

Rice

Milk

Raisins

This recipe may be sufficient for someone who has made the dish many times and may simply serve as a reminder to ensure that all the proper ingredients are included. The problem is that the recipe does not give nearly enough information or instruction for another person to duplicate it. In preparing a recipe, the goal of foodservice operations, unlike that of home cooking, is to ensure consistency.

Foodservice operations can obtain recipes from a multitude of sources: books, magazines, food product manufacturers, and so forth. The recipes obtained from these sources may have been compiled by professionals, but they still are not standardized recipes. A **standardized recipe** is one that has been tested and adapted to the equipment and staff of a particular operation. A recipe that is standardized for one operation is not standardized for another. A foodservice operation must produce the recipe several times in order to ensure the quality of the product and to see that the recipe yields the expected number of portions. These steps ensure that there will not be any surprises in the production of the menu items, that the cooks have an opportunity to practice on the items and adjust the recipes, and that the service staff is given an opportunity to sample the menu items.

When designing and developing a standardized recipe one should assume that the person who will be using it, making the menu item, has never made the item before and may have only limited food-preparation experience. The greater the amount of information included, the more easily people can use the recipe properly. The skill level of employees fluctuates greatly, from barely experienced to highly experienced; it is important to make sure that even a minimally experienced cook can follow the set of instructions so that the resulting item will be as intended. The following paragraphs discuss the information a recipe should include in order to produce a consistent product time after time.

THE IMPORTANCE OF USING STANDARDIZED RECIPES

A recipe, as mentioned earlier, is a set of instructions and ingredients used to make a certain food or beverage item. The more detailed the set of instructions and the list of ingredients, the greater the possibility that the item will be produced consistently from time to time. It is important for foodservice operations to produce and develop standardized recipes for their operations so as to control food costs and to maintain product consistency.

1. *Training.* Developing a written set of recipes makes the task of training kitchen personnel much simpler. A written recipe also provides a guide for the cook to follow in the preparation of the dish when being trained, as well as a source to refer to each time the dish is prepared subsequently. A problem may occur in operations that do not have written recipes: A new cook is often trained by several different people, each of whom shows, the cook a different way to prepare an item. The result is a confused cook and the potential for an inconsistent product. Without a written recipe, the new cook does not know which is the proper way to make the dish.

2. *Ordering.* An accurate and detailed list of ingredients for the menu items provides valuable information for the purchaser of the operation. The purchaser can simply order supplies based on the amounts needed to serve the number of portions forecasted.

3. *Standardization.* Information that is passed verbally from person to person can become distorted. The same is true of recipes that are not written down. Foodservice operations must have written recipes to ensure that items are the same each time they are made.

Key Components of a Standardized Recipe

1. *Name of the Recipe.* People must know what to refer to.

2. *Yield.* Total yield, or quantity of food the recipe produces. A recipe should also include the number of portions and portion size (for example, 50 @ 6-ounce portions).

3. *Ingredients.* Ingredients should be listed with specific measurements in the order in which they are needed in the recipe.

4. *Equipment.* All pieces of equipment needed to produce the item should be listed.

5. *Specific Directions*. Directions must be sufficiently detailed that introductory-level personnel can understand them.

6. *Times for Cooking*. The amount of cooking time the dish requires should be included as a guideline. The inclusion of preparation time is also helpful to assist in planning.

DEVELOPING STANDARDIZED RECIPES

Recipes must provide the information necessary for foodservice workers to produce menu items consistently. In developing a standardized recipe, it is important to follow these guidelines:

1. *Accurate Descriptions of Ingredients*. The recipe must include accurate descriptions of the ingredients. Food items such as apples, flour, and beef come in many different forms and types that can influence the outcome of the recipe. The more specific the description of a food item, the better the chance that the recipe will yield consistent results.

2. *Accurate Ingredient Amounts*. The recipe must include clear and accurate ingredient amounts. Measurements given in weight rather than volume are more accurate. Ingredient amounts listed as "pinch," "dash," or "to taste" make it difficult for cooks to produce the recipe consistently. Such vague amounts may also confuse a cook who has never made the dish before.

3. *Accurate Yields*. Recipes must state how much of the item it will make. This information is crucial to ensure that there will be enough of the item, but not too much. Only the amount of food that is needed to handle the forecasted business should be prepared.

4. *Specific and Detailed Set of Instructions*. Preparation personnel need a set of instructions detailed enough for a novice cook to understand. Steps must be listed in sequential order for ease of use. Instructions such as "Cook until done," without a description of what exactly "done" is, could be misleading to those who have not made the dish before.

Standardized recipes do not guarantee food consistency or improve the training of new kitchen perparation personnel. A recipe is only as good as the effort put into ensuring that it is used properly; a recipe that is never used does the operation no good. Cooks must be trained in the proper way to use each recipe and must be provided with the necessary products to produce the items. Management should make sure that kitchen personnel are using and following the operation's standardized recipes.

CONVERTING RECIPES

The number of portions, or **yield,** of a recipe differs according to its market or intended audience. Recipes written for home or noncommercial use generally yield only about 4 to 6 portions, whereas those for commercial use are generally written for either 20 or 50 portions. Regardless of the size of a recipe's yield, it rarely produces the exact number of portions needed by the preparation personnel. Most recipes must be converted, or changed, to fit the yield that is needed. A recipe must be accurately converted so that food and labor are not wasted. If the cook does not make the necessary amount of food, the operation could run out of servings, which could cause problems with customers.

The conversion of a recipe from the **given yield,** (that of the original recipe) to a **new yield** (the number of portions needed by the operation at a specific time) is simply the matter of finding the ratio, which is the **conversion factor,** of two numbers: the new yield and the given yield. Once the conversion factor is calculated, it is multiplied by the ingredient amounts in the original recipe to determine the ingredient amounts needed to produce the new yield. The procedure is the same for both increasing and decreasing the yield of the original recipe.

A guideline to remember is that if you need to increase the yield of the initial recipe, you will need a conversion of greater then one; if you need to reduce the yield of the original recipe, you will need a conversion factor of less then one.

FORMULA FOR CONVERTING RECIPES

Conversion factor *(CF)* = new yield/given yield.

Multiply *CF* by ingredient amounts in original recipe.

CF × ingredient amounts in original recipe = ingredient amounts to produce the new yield.

Remember: To increase the yield of the original recipe: CF > 1.
  To decrease the yield of the original recipe: CF < 1.

For example, the recipe for Orange Cucumber Salad is as follows:

Orange Cucumber Salad

Yield: 4 portions

Ingredients

2 oranges

2 medium cucumbers

1 small red onion

1 cup Orange Dressing

1 head romaine lettuce

Procedure: In converting or changing the yield of a recipe, the procedure or list of steps does not change. The yield for the original recipe is 4, but 20 salads are needed.

New yield = 20   Original yield = 4
Conversion factor = 20/4 = 5

Multiply all of the ingredient amounts in the original recipe by the conversion factor to determine the ingredient amounts for the new yield of the recipe.

Orange Cucumber Salad

Yield: 20 portions

Ingredients

$5 \times 2 = 10$ oranges

$5 \times 10 = 50$ medium cucumbers

$5 \times 5 = 25$ small red onions

$5 \times 5 = 25$ cups Orange Dressing

$5 \times 5 = 25$ heads romaine lettuce

YIELD TESTS/QUALITY CHECKS

The process of standardizing a recipe for a particular operation involves making the product several times to be certain that it is consistently the same and to allow for adjustments. This also gives the kitchen staff an opportunity to practice and to become more proficient at making the food item.

The repetition of using the recipe also allows the kitchen staff to determine that it produces the proper yield or number of portions. The **yield test** is crucial for cost control. The management staff must have an accurate idea of the yield so that it can produce the proper amount of a food product.

While checking to ensure that a recipe produces the stated number of portions, the kitchen operation also has an opportunity to perform a **quality check** on the item. Sample groups should be allowed to taste the food item and comment on its acceptability. Groups of both front-of-the-house and back-of-the-house employees should be consulted. Once an item has been approved by these groups, a sample group of customers can be consulted. Some operations offer a new food item as a special in order to gauge customer comments in regard to its permanent inclusion on the menu.

DETERMINING THE SELLING PRICE OF MENU ITEMS

Setting the selling price for the menu items in a foodservice operation is essential to the financial success of the business. Management must calculate a selling price that covers not only the cost of the food item but also the other costs of the operation plus a profit.

The first step is to determine the accurate cost of the menu item. This process is made simpler with the use of a standardized recipe. The recipe provides a list of all ingredients used to make the item, an accurate indication of the quantity of the ingredients, and the number of portions that the recipe yields.

Recipes can be costed either manually or with the aid of a computer. The calculated cost of the recipe is good only insofar as ingredient costs are accurate. Once the cost of an ingredient changes, which is common in dealing with fresh food items that fluctuate in price, the accuracy of that item's cost diminishes.

The reduction in the cost of computer hardware and software has made computers more accessible to a wider range of operations. Food management computer software stores information on the inventory of the operation and the standardized recipes. The advantage of using a computer is that once the information is stored, product prices may be easily updated to determine accurate recipe costs.

The process of costing out recipes manually, or without a computer, is often a long and tedious one. The person responsible for costing out a recipe calculates the cost of each ingredient to determine the total recipe cost, then divides that by the number of portions the recipe makes to determine the **cost per portion.** The problem is that if the price of any of the ingredients changes, the calculated cost becomes inaccurate.

Calculation of a Recipe's Cost per Portion

Cost of ingredient 1

+

Cost of ingredient 2

+

Cost of ingredient 3

+

Cost of the rest of the ingredients

Total cost of the recipe ingredients

Cost per recipe portion (CPP) = total cost of the recipe ingredients/yield of the recipe

Meals that are served to guests are generally made up of several items. For example, a breakfast plate may include a cheese omelet, hash browns, and toast. Although all these items served to the guest are included in one price on the menu, the cost of each recipe or item must be calculated separately and then totaled to arrive at a plate cost for the dish.

Once the plate costs are calculated, the manager can use them to determine the selling price of the meal. Management must be careful to ensure that the operation charges enough for the menu items to recover all costs of the operation, as well as to provide a profit for the business. It is also important for an operation not to charge too much for what it sells, so as not to price itself out of the market or to charge more than its customers are willing or able to pay.

THE FACTORS TO CONSIDER IN SETTING SELLING PRICES

Determining the selling prices of menu items requires careful consideration. If an operation sells a menu item taking into account only the cost of the food, it will lose money. The selling price must not only reflect the cost of the food, but must also provide money to cover the costs of the business, including profit.

A problem in determining what to charge for a menu item is the lack of an accurate determination of overhead and the other indirect costs involved in making the item. It is difficult to determine accurately what part of these costs can be allocated to any one item an operation produces. The cost of the food that goes into any item is relatively simple because the exact ingredients are known. But it is difficult to calculate the cost of the utilities of the operation (gas, electricity, and water) used to make a particular dish, or an accurate measure of the cook's time per dish. Although management knows the hourly cost of the cooks that prepare the food, very seldom does a cook work on preparing one food item at a time. A cook's time is spent in preparing a number of items at once, inasmuch as a portion of the preparation time for most food items is the time the food is cooking, which does not require the cook's undivided attention. For example, while a roast is in the oven, the cook can devote time to preparing other items. The inability to allocate accurately the cost of rent, utilities, and labor to specific menu items complicates the determination of a selling price.

The setting of menu prices is no easy task. There are many **pricing factors,** both tangible and intangible, that must be considered. Tangible factors, such as the costs of the food items, labor, and overhead are essential in determining the price at which to sell an item. There are also intangible factors, such as management's perception of the value of the item, what it feels the item is worth, and the prices competitors are charging. Such intangible factors can have a significant effect on the price of a menu item.

Some segments of the foodservice industry, such as fast food, are very price sensitive. Customers of this industry segment will switch their business from one operation to another based solely on a small reduction in price. Management must be aware of this fact and take it into account when setting prices. For example, an operation may calculate an ideal price, using all the tangible factors discussed earlier, to ensure that all costs are covered. Then the local competition runs a special on an item that both operations sell and drops the price slightly. Unless the first operation reacts, it runs the risk of its customers going to the competition.

EXAMPLES OF METHODS FOR CALCULATING MENU PRICES

There are a number of methods that foodservice operations can use to calculate the selling prices of menu items, taking into account the limitations in cost allocation. Both methods discussed here determine selling price based on tangible factors. Management then needs to adjust this price according to the intangible factors to decide what the operation will charge. A premise of both methods is the use of estimations to determine the costs.

Multiplier Method

The **multiplier method** is one of the most commonly used in foodservice operations owing to its ease of use. The management of an operation determines a multiplier, or a number by which it will multiply the basic menu item costs in order to cover the costs of the operation. The multiplier, although somewhat arbitrary, depends on what the operator thinks the operation will need to cover the rest of the expenses of the business and to provide a profit.

Management determines what it would like to use for a food cost percentage for the operation—usually 33% to 50%. To determine the multiplier, the food cost percentage is divided into 100. For a desired 33% food cost percentage, .33 divided into 100 yields a multiplier of 3. The cost of menu items is then multiplied by 3 to determine how much to charge on the menu.

MULTIPLIER METHOD

Desired food cost % / 100 = **Multiplier**

Multiplier × cost of menu item or plate cost = **Selling price**

Once the multiplier is determined, it can easily be used by those who calculate the selling prices. This method, however, has its negative aspects. Its simplicity of use must be balanced with the inaccuracy of the calculation of some items. This method marks up all items the same amount, regardless of any other factors that could raise the cost of an item, such as labor. The cost of an item that requires a lot of prep work, such as lasagna made from scratch, is higher than the cost of items that require minimum preparation time and labor, such as a hamburger that is purchased frozen and requires only cooking and assembly. Both would be marked up the same amount even though they require very different amounts of labor.

The Texas Restaurant Association Method

The **Texas Restaurant Association method** is considered to be the most accurate in determining the selling price of menu items. Rather than lumping into one category all the food items an operation sells, regardless of the amount of labor required to produce them, this method separates menu items into two categories. One category includes items that require a great amount of labor, such as those an operation makes from scratch. The second category comprises items that do not require much labor to produce, such as those that are purchased ready to cook or serve without much use of a cook's labor—for instance, hamburgers or steaks that are simply taken from a box and cooked.

The increased accuracy of this method is achieved by dividing food items into the two categories according to the amount of labor required to prepare them. Grouping items this way allows a better and more equal split of the costs, resulting in a more accurate determination of menu prices.

OTHER FACTORS TO CONSIDER IN DETERMINING MENU PRICES

The preceding methods are examples of the various ways in which operations can determine how much to charge for their menu items. These methods provide a dollar figure that should serve as a guideline for management rather than dictate a definite price.

It is also important to examine the gross profit that a menu price contributes to the operation. **Gross profit** is the amount of money remaining once the cost of ingredients is subtracted from the sales price of the item.

Gross profit = $ sales price – cost of ingredients

The expenses of the operation (overhead, labor, etc.) are paid from the gross profit. After all those expenses are paid, any dollar amount that is left over is the profit for the operation. Because of their high costs, it may not be possible to sell some menu items, such as lobster or larger steaks, at a high enough price to derive the cost percentages that the operation would prefer, based on one of the two methods discussed earlier. Although the selling price would have to be reduced in order to sell it, the item would still contribute a significant gross profit to the operation. Consider the following example:

Maine Lobster Dinner		*Fried Chicken Dinner*	
Salad/dressing	.27	Salad/dressing	.27
Roll/butter	.11	Roll/butter	.11
Fresh lobster	10.00	Fried chicken	1.25
Vegetable/starch	.22	Vegetable/starch	.22
TOTAL	$10.60	TOTAL	$1.85

If the operation uses a price factor of 3 for a 33% food cost %:

Menu sales price = pricing factor × total cost per plate

Lobster $31.80 = 3 × $10.60
Fried chicken $5.55 = 3 × $1.85   Multiplier

At $31.80 for the lobster, management keeps the cost percentage calculated to cover the rest of the costs of the operation, but the price may be too high for the restaurant's customers to afford and, therefore, they will not order it. If customers do not order the item, the operation loses both the sale and the cost of the lobster. The selling price for the chicken, $5.55, may be a little low, depending on the operation, and may warrant an increase to keep the price in line with the other items the operation sells. Management must consider all these factors when setting menu prices.

In this case, management would be wise to lower the menu price of the lobster to $25.00 to attract more sales, while still relying on the high gross profit that the dish will contribute to the operation. The price of the chicken may be raised to $7.00.

Gross profit = sales price – cost of ingredients

Lobster $14.40 = 25.00 – 10.60
Chicken $5.15 = 7.00 – 1.85

The lower price for the lobster dinner will stimulate sales of the item, while still providing a significant amount in gross profit to the operation. Although the chicken contributes a significantly higher percentage of its sales price to the gross profit, the operation will have to sell more than 2.5 chicken dinners to make the same gross profit as it will with the sale of one lobster dinner.

The difficulty for a foodservice operator is to set menu prices so that the costs of the operation will be covered and an adequate profit will be provided for the business. If the operator charges more than the customer is willing to pay, or more than the customer feels the item is worth, the customer will not patronize the business. If the operation does not adequately determine the costs of a menu item and sells it for less than the cost to produce it, the customer will enjoy it and return, but the foodservice operation will soon be out of business. The menu price, as determined by one of the many methods available to a foodservice operator, should be considered only a guideline and not a definitive price. Other factors must be considered in determining the selling price of an item, which may raise or lower the price.

Customers often have a preconceived notion of what a menu item is worth. The perceived value of an item can depend on what others charge for it. When a foodservice operation sells an item at, or below, the customer's perception of its value the item will sell better than an item the customer perceives as selling for more than it is worth.

Besides taking into account what the customers may be willing to pay for an item, the management must consider other factors as well. The prices that an operation's competition charges for a similar item can be of considerable importance to management in determining selling prices, especially in operations whose market is made up of people who are price sensitive, or who use price as one of the main determinants when deciding where to dine. Fast-food restaurants serve this type of clientele; in this market a $.25 difference in cost can draw a customer from one place to antother.

THE USE OF COMPUTERS IN FOOD-COST MANAGEMENT

Computers have simplified the task of monitoring and controlling food costs in a foodservice operation. This, coupled with the decline in computer hardware and software prices has increased the availability of this important tool to many operations.

HUMAN RESOURCES AND LABOR COST MANAGEMENT

The cost of labor for foodservice operations has continued to spiral upward for a number of reasons. Foodservice operations are labor-intensive by nature, requiring many labor hours to accomplish the tasks that need to be done. Customers want to be served by a person, not by a machine. The tasks performed in foodservice operations, such as making salads and dicing vegetables, are generally done to produce small quantities, so doing these jobs with a machine may not prove to be an efficient use of time, once the time required to set up and clean the machine is calculated.

The erratic nature of the foodservice business also contributes to high labor costs. Most operations do not know exactly how many people will come in to eat, or when during a meal period these customers will arrive. This makes scheduling and labor control a substantial challenge. Management must be careful not to over-schedule and thus waste labor dollars, nor to underschedule and risk leaving customers underserved.

Labor at one time was both cheap and plentiful for foodservice operations. This situation changed as the labor market tightened and the government raised the federal **minimum wage**. Foodservice, an industry that relied heavily on the abundance and availability of cheap labor, was generally caught off guard and forced to find alternatives to the shrinking and increasingly expensive labor market. This situation, coupled with a high rate of **turnover,** or the number of people that had to be hired to fill one position within a year's time, raised costs significantly. The repetitious nature of many foodservice positions, the low pay, and the sometimes less-than-ideal working conditions all contribute to high turnover. Whenever an employee quits and another must be hired to take his or her place, there are additional labor costs for the operation because of the expense of hiring and training the new worker.

The first step in controlling labor costs is to institute an effective system to manage the human resources of the operation. Management must understand that the workers of the operation are an essential resource and that they have a strong influence on the effectiveness and success of the business. The process of recruiting, hiring, and training employees is an important aspect of human resource management and a crucial component of labor cost control.

EMPLOYEE RECRUITMENT

The first step in the process of human resource management is the recruiting of employees to staff the operation. The more effective the recruitment, the more successful the whole operation. Advertisements for vacant positions should be posted in places that are visible to the workers the operation seeks to hire, for example, in local newspapers or trade association publications. Another good source of potential employees is the operation's current employees. Some operators offer a small cash bonus for referrals by employees.

Checking the professional references of an applicant is essential. No one should be hired without management's talking to his or her former employers. Too many dishonest employees are able to move from one job to another without their past problems being detected.

The recruitment process is often made difficult by the employee who does not give a two-week notice when he or she decides to quit. Without a sufficient amount of time, the process of attracting new employees is more difficult. Some managers, forced to fill a position quickly, resort to the not-recommended process of back-door hiring. **Back-door hiring** describes the situation when a potential employee appears at the operation inquiring about a job and is hired without filling out an application and without having his or her work experience or references checked. The process fills the vacant position, but may come back to haunt management if the employee causes problems.

Consider the following example of a problem caused by back-door hiring. A worker applied at a seafood restaurant the day an employee quit without giving notice. The person was hired on the spot and began to work the lunch shift at the deep-fat fryer station. The employee said that he had worked at a seafood place across town, when actually he had not. It was his wife who worked at the restaurant across town, and who told him what he needed to know about how to deep fry food. The manager, in a rush to hire someone to replace the employee who left without notice, had felt that there was neither the time nor the need to call any of this worker's past employers. The new employee did an adequate job frying the food. When the meal period ended, the manager shut off the fryer and asked the new hire to drain and clean the fryer. The employee got a plastic bucket and began to drain the oil into it. The hot oil melted the bucket and the employee's shoes, causing permanent damage to his feet and costing the restaurant considerable medical and legal expense. This example, although extreme, illustrates the risk in not calling to check a person's work history.

JOB ANALYSIS AND JOB SPECIFICATION

A **job analysis,** or description of the work responsibilities of the position, must be developed for each position in an operation. This valuable tool includes the equipment a person must be able to operate to fulfill the requirements of the position. The **job specification** is another component of the job analysis. The specification lists the educational training, prior work experience (if any), and any physical capacities that a worker must possess to do the job adequately.

Compiling a job analysis is a time-consuming task, but one that will pay off in the long run. Writing up the responsibilities and requirements for all the various positions of the operation provides the material for the job descriptions that are given to employees, ensuring that there is no misunderstanding between them and management about what their jobs require.

PERFORMANCE APPRAISAL

Employees should be made aware of their performance and how they are measuring up to the standards set for them. A **performance appraisal** should be conducted at regular intervals and should include what the employee is doing right and which areas need improvement. The appraisal of an employee is often prefaced by the employee's completing a self-evaluation. To be most effective, an appraisal should not surprise employees with infractions or problems that occurred in the past but were never mentioned. The entire process should be straightforward, with the employee knowing of the appraisal in advance, knowing what is expected of him or her on the job, and being aware of the points the evaluation will cover. The employee who is comfortable in the job and knows what is expected of him or her will be more productive than one who does not.

TRAINING

The key to an effective and productive staff is training. Most of the turnover of workers in the first couple of weeks of employment can be attributed to a lack of training in their jobs. Without proper training, management cannot expect a new worker to perform his or her job well. Even though an employee may have had plenty of experience in other jobs or foodservice operations, he or she must be trained in the new operation's standards and operating procedures. It is extremely important not to rely on an employee's word that he or she knows how to perform a certain task. The multitude of dangerous equipment and situations in a foodservice operation requires that management make sure an employee knows how to perform a particular task before being allowed to actually do that task.

Employee training, although crucial to the operation of any foodservice, is sometimes neglected by employers, thinking that it is too expensive or time-consuming. Of course, training is expensive and requires an investment up front, yet it will pay off many times over in the long run. Training materials should be both written and formalized and should parallel the materials giving the position's job description and analysis. By supplying employees with written materials, an operation can ensure that the training is consistent and that management has a record of what was covered.

Employees must be trained in both the operation's methods and its management policies. The more aware an employee is, of how the operation functions and what it stands for, the more effective that employee will be. Each employee should be given initial training at the beginning of employment, as well as continued training throughout his or her time with the operation.

SUMMARY

Foodservice operators must be aware of the costs necessary to operate their businesses. There are many opportunities for costs to get out of control, causing a business to suffer financially. The low profit percentage of foodservice operations, 10% to 20%, does not leave much room for error. The more managers understand about the nature of the various costs, the better they can control them. Although some increases in costs are beyond a manager's control, there are many things he or she can do to control costs. The use of standardized recipes and accurate menu item costs and selling prices gives management a handle on maintaining food costs.

Employees are integral to the success of any foodservice operation. They provide the personal service that draws customers. Foodservice operations must develop and use an effective system of managing their human resources. Careful planning is essential in all aspects of human resource management, starting with employee recruitment and continuing through job analysis and specification, performance appraisal, and training.

KEY TERMS

fixed costs	pricing factors
variable costs	multiplier method
overhead costs	Texas Restaurant
recipe	Association method
standardized recipe	gross profit
yield	minimum wage
given yield	turnover
new yield	back-door hiring
conversion factor	job analysis
yield test	job specification
quality check	performance
cost per portion	appraisal

DISCUSSION AND REVIEW QUESTIONS

(Choose any 4)

1. Why is cost management so important to the success of a foodservice operation?

2. Name the three main categories of costs in foodservice operation? List three key components of each of the three main categories of costs.

3. Name four activities that occur in a foodservice operation that could cause a rise in the food costs of the operation. Name three things that make controlling food costs difficult.

4. What is the difference between a recipe and a standardized recipe? Name two reasons for the importance of standardized recipes. Describe three of the key components of standardized recipes. Do standardized recipes guarantee food consistency? Explain.

5. What actually is being converted when one converts a recipe?

6. Why must foodservice personnel know how to convert recipes?

7. Why are yield tests and quality checks needed? What does each measure?

8. Name some of the problems that make it difficult to determine accurately the cost of menu items. Explain.

9. If an operation determines the food cost of its hamburger lunch plate is $1.25 and decides to sell the lunch plate for $1.25, is it making a profit, losing money, or breaking even (no profit/no loss)? Explain.

10. State the two methods for determining menu selling prices. Name an advantage and a disadvantage of each.

11. What does "labor-intensive" mean? Why are foodservice operations labor-intensive? What aspects of the foodservice industry help to increase labor costs?

12. Suppose that the manager of a foodservice operation gets a call from one of the cooks at 3:00 P.M., saying that she is quitting and will not be in to work for her scheduled shift that evening. The manager is expecting a very busy night and does not have another cook to work the dinner shift. At the same time, a person comes into the restaurant looking for a cook's job; the manager hires the person on the spot and puts him to work. What is this hiring practice called? Is it a wise practice? Explain. Describe two potential problems associated with this type of hiring.

13. What is the difference between a job analysis and a job specification?

14. What is the key to having an effective and productive staff? Explain why this key is so important.

15. Modguual of F.S.O multi paragraph essay on what I learned the most/least

Due on Monday 11th

GLOSSARY

À la carte menu: A menu for which each item is priced separately.

Accident: Random occurrences that cannot be avoided.

Accommodation: The goal of the server and kitchen personnel to do whatever it takes to satisfy guests by accommodating their desires.

Accountable: Responsible for specific jobs, such as cleaning.

Aerobes: Bacteria that need oxygen to live.

American plan: A form of payment for a hotel or resort in which meals are included in the cost of the lodging.

American service: A style of service in which the food is cooked and plated in the kitchen and the server simply places the finished plate in front of the guest.

Anaerobes: Bacteria that do not require oxygen to survive.

Anticipation: Servers who anticipate their guests' needs are more effective and can serve their guests better.

Attentiveness: Guests often dine in a restaurant rather than eat at home because they want someone to attend to their needs. An attentive server will acknowledge guests as soon as they arrive, refill water glasses when low without being asked, etc.

Attitude: The attitude of the server has a dramatic effect on a customer's dining enjoyment.

Average check: The average amount of money a guest spends at a foodservice operation.

Back-door hiring: Filling vacant positions without the filling out of applications or checking of experience, qualifications, or references.

Back of the house: A common phrase to describe the kitchen.

Bottom line: A term used for profit, referring to the fact that profit is on the "bottom line" of an income statement.

Brand name: Designation of a product by the name of its producer or manufacturer.

Broiler: A cooking unit, heated either from above or below, used to cook individual portions of meat, poultry, or seafood.

Broken case: A term used in purchasing when a supplier sells less than a full case of goods.

Buffet service: The category of self-service in which customers choose what they wish to eat from a selection displayed on a table or counter.

Buying: A process to obtain goods that is less complex than formal purchasing, in which specifications are not used; generally, only one supplier is called and prices between suppliers are not checked.

Cafeteria service: Self-service used to feed large numbers of people in a short period of time with a minimum of labor.

Captive audience: A term used to describe customers who either have already paid for their meal or are unable to dine elsewhere.

Carbohydrates: Primarily sugars and starches found in foods derived from plants.

CDC: Center for Disease Control.

Changing menu: A menu that changes due to one of several reasons; the opposite of a static menu.

Chef de rang: The head server in the two-server French system.

Cholesterol: A soft waxy substance manufactured by the body.

Classical dishes: High-quality menu items prepared according to traditional methods and under high standards of excellence.

Classical school: A study of management practices that complemented the scientific management method by developing a fixed set of principles that established the basis for management.

Clean: Lack of visible dirt or of items that generally do not belong.

Clean-as-you-go: A system in which employees are held responsible for cleaning the areas where they work and the equipment that they use.

Combination menu: A type of menu that combines several individual types (e.g., table d'hôte and à la carte).

Commercial, for profit: Operations that are run to make money for their investors or owners.

Commis de rang: The second or assistant server in the two-server French system.

Commissary: A foodservice operation that produces food items to be served in another location.

Committed patron: A foodservice customer who orders only menu items considered healthy when he or she dines out.

Compartment steamer: A metal box with steam injected that is used to cook and reheat food.

Concept: The theme of a foodservice operation.

Conflict of interest: When buyers benefit personally from their dealings with the operation for which they work.

Consensus: Agreement reached by a group of people.

Contamination: The inadvertent presence of bacteria or harmful substances in food, beverages, or water.

Contingency theory: A management theory that states that there is not one single best way to manage people or an organizational structure for all situations or types of business.

Convection oven: An oven that uses a fan to circulate the air, resulting in reduced cooking times.

Convenience food system: A foodservice operation that does not produce the items it serves but, rather, serves already prepared items purchased from a supplier.

Conventional oven: The most basic form of oven.

Conventional/traditional foodservice system: A system in which food is served to the guest as soon as it is prepared.

Conversion factor: The ratio of *new yield* to the *given yield* of a recipe.

Cost per portion: The total cost of all ingredients in a recipe divided by the number of portions produced.

Critical control point: A procedure of a recipe in which a preventative measure can be applied that would either eliminate or minimize a hazard.

Cross-contamination: When one product or utensil spreads bacteria to another, allowing harmful substances to come in contact with new products.

Customer: Someone who exchanges his or her money for goods and services.

Customer-driven: Designing an operation in which the needs of the guests drive the decisions.

Cyclical menu: A menu that changes on a regular or set basis.

Demographics: Information such as age, income level, number of children, and so on, used to group people into categories for marketing purposes.

Descriptive copy: The written description of a menu item used to inform the guest and help sell the item.

Dietary fiber: Carbohydrates found in fruits and vegetables that cannot be broken down by the human digestive track and are beneficial because they promote the feeling of fullness.

Dough hook: A mixer attachment consisting of a large bar, bent into an S-shape, used for mixing doughs and other heavy-consistency substances.

Du jour: A menu item that changes on a daily basis.

English service: Also called host or family service, is less common in commercial foodservice operations in this country. The food is prepared in the kitchen, brought to the dining room in bowls and on platters, and portioned and plated at the table by the "host."

Entrepreneur: A person in business for himself or herself rather than working for someone else.

European plan: A method of payment for a hotel or resort in which the guests pay for their lodging and meals separately.

Facultative: Bacteria that can survive with or without oxygen.

Family service: Type of service in which food is brought to the table in serving bowls and platters, allowing guests to help themselves.

Fast-food: A high-volume type of commercial restaurant that provides food quickly and inexpensively by offering limited menu choices and reducing production and labor requirements.

Fat: The body's main form of storage. Fat from animal sources is generally not good for the human body, while fat derived from plants is better.

FDA: Food and Drug Administration.

FDA Model Sanitation Ordinance: A set of federally developed standards, used as the basis for state and local health department food regulations.

FIFO: An abbreviation for First In, First Out—an inventory rotation system in which products on the shelf the longest are used first.

Fixed costs: Costs that remain fixed or constant, regardless of the amount of business (e.g., rent, salaries, and insurance).

Flat paddle/beater: A mixer attachment with one broad, flat blade, usually used for making batters, mashing potatoes, and creaming butter.

Follow-up: In cleaning programs, inspection of areas by managers after each shift, to ensure that established procedures are being followed, with employee evaluations and rewards.

Food chopper: A tool used to cut and dice vegetables and other food items.

Food contractor: A company that is hired by another company, school, college, or state or federal government to run its foodservice operations for them.

Food court: An area that is lined with several different foodservice operators sharing a common dining space.

Food-borne illness: An illness that results from eating food containing live bacteria that can cause illness.

Food-borne illness outbreak: A reported incident in which two or more people become ill from a common food, confirmed by laboratory analysis.

Food slicer: A motor-driven metal disk with a sharp edge, used to cut, uniformly and accurately, meats, vegetables, and other foods suitable for slicing.

Formal purchasing: The process of purchasing, using written specifications, purchase orders, and so on.

Franchise: A formal agreement between two parties in which the *franchisee* agrees to run an operation using the standards and procedures of the *franchiser* company.

Free-standing broiler: A broiler whose heat element is under a bed of ceramic coals to simulate charcoal cooking, the food being held on a grill so that the fat drains through.

Free-to-choose: Customers who have several options of where to dine and who to pay by the meal.

French service: A service system in which the completion of the cooking, portioning, and plating is done in the dining room utilizing the two-server system.

Fresh: A product that has never been frozen, canned, or dried.

Front of the house: Another term for the dining room.

Full-service restaurant: A foodservice operation that has a full choice of selections on its menu and offers table service.

Generic product: A product that does not have a brand name or a name that is recognized by customers.

Gross profit: The amount of the selling price that remains after subtracting the cost of ingredients.

Guerdon: The cart food is cooked, portioned, and plated in the dining room for French service.

HACCP: Hazard Analysis Critical Control Point, a food safety inspection system that highlights potentially hazardous food.

Hand truck: A two-or four-wheeled cart used to move food and equipment from the receiving area to where they are needed.

Homemade: Prepared on the premises, not purchased already prepared.

Human relations school: A school of management theory that first looked at workers as humans with feelings rather than as machines.

Implementation: Setting up measures to ensure that employees understand and carry out a program, such as a master cleaning schedule.

Independent ownership: A foodservice operation owned by one person or a group of people and not affiliated with other restaurants.

Informal buying: See buying.

Infrared oven: An oven that cooks using radiant heat, used for heating and reheating small items.

Intangible: Something that cannot be touched or handled, such as service.

Invoice: A document used as a sales receipt for goods purchased and delivered by a distributor to a business.

Job analysis: A description of the work responsibilities of a position in an operation.

Job specification: A component of job analysis that lists the educational training required.

Kickback: An illegal ploy of purchasing agents in which they require that suppliers give them money or goods personally in exchange for an order.

Labor intensive: Something that takes much labor to accomplish.

Lead time: A term in purchasing meaning the amount of time between when items are ordered and when they will be received.

Limited menu: A menu that offers guests limited types of food, such as pizza, chicken, or sandwiches.

Low-ball pricing: A process in which a supplier quotes a low price, enabling a foodservice operator to get or keep his or her business, but then the supplier is forced to raise the price of other items to make up for it.

Manager: The role of the manager is to ensure the smooth operation of a foodservice establishment. Managers are responsible for making the decisions that steer the operation in the right direction.

Market: The general environment in which business transactions take place.

Market research: An investigation undertaken to determine what an operation must offer to satisfy its customer base.

Marketing tool: A means of attracting, communicating to, informing, and persuading customers to buy.

Market share: The percentage of the business your operation has.

Master cleaning schedule: A detailed list for a cleaning program for a foodservice operation that includes the following four components: what is to be cleaned, when it should be cleaned, whose job it is to clean it, and when the cleaning should be completed.

Master distributor: A foodservice supplier who handles a wide range of suppliers, allowing a foodservice operation to purchase all of their supplies from it if desired.

Meal patterns: The number of guests and rate they arrive at a foodservice operation.

Mechanical oven: An oven in which the trays where the food is to be cooked are moved to produce even cooking of the product.

Merchandising: The process of making an item more attractive to customers so that they will purchase it.

Merchandising tool: A menu is an effective merchandising tool; descriptions of menu items along with pictures help sway guests to order certain items.

Microwave oven: An enclosed box in which waves are generated to vibrate food and water molecules, causing heat through friction.

Middle-level managers: Positions such as bar or restaurant manager, whose role is to monitor and control the daily operation of the business.

Minimum wage: The lowest allowable per-hour rate of income, set by the federal government.

Mixer: A tool used for combining ingredients, beating, whipping, folding, and so on.

Multiplier method: A method of menu item pricing in which the management determines a number by which it will multiply the basic costs in order to cover all costs of the operation.

Noncommercial: An operation that is not run for profit.

NRA: National Restaurant Association.

Nutrients: The building blocks that cells use to grow and maintain themselves.

Nutrition: The study of foods and their relationship to health.

Nutritionist: A professional who has studied nutrition and has been certified for his or her expertise in the relationship between diet and health.

Ordering: The process of relaying to the supplier the items one needs.

Organization: A group of people united to achieve common objectives or goals.

OSHA: The Occupational Safety and Health Administration, the federal regulatory agency in charge of enforcing safety and health standards of American workers on the job.

Oven: An insulated box with a heating element used to cook food.

Overhead costs: Costs of running a business that are not food or labor related.

Perceived value: The comparison between what a customer pays for a meal and what he or she thinks the meal is worth.

Performance appraisal: A periodic evaluation of how an employee measures up to preestablished standards for carrying out his or her job.

pH: A measure of the alkalinity or acidity of a product.

Physical hazard: A stone, a staple, or any other item that inadvertently gets into food.

Point of origin: The geographical location from which a food item is obtained.

Portioning: Separating a food item in large quantity into the quantity it will be served to a guest.

Potentially hazardous food: Food that, due to its composition, is more susceptible to bacterial problems and should be handled with care in all phases of the foodservice operation.

Prepared "in-house": (Food) produced entirely from the unprocessed state within the operation.

Pricing factors: Tangible and intangible components in menu prices, including food, labor, overhead, perceived value, competing prices.

Proactive: Taking a stand on an issue and planning to avert a problem before it becomes one.

Production personnel: Another name for the kitchen staff.

Protein: Serves several purposes in the human system, but its primary function is to help the growth and maintenance of cells that make up the body.

Purchasing: A systematic and planned process of determining what is needed, checking prices, negotiating with suppliers, and obtaining the goods needed for a foodservice operation.

Quality check: Tasting of food items by foodservice employees to determine and comment on their acceptability.

Quick service: Another name for a fast-food restaurant operation.

Range: A cooking unit, heated by gas or electricity, designed to cook food in containers as well as on a flat-topped grill.

Reach-in refrigerator/freezer: An open storage unit, smaller than a walk-in unit, usually placed closer to the production line and the meal service area.

Ready-food system: A system developed to smooth the fluctuations in a foodservice operation by having production personnel (cooks) prepare menu items in large batches that are then portioned into serving sizes and packaged, to be reheated and served later.

Receiving: The process by which the ownership of the items purchased changes from the food distributor to the foodservice operation.

Rechaud: The cooking apparatus, placed on a guerdon, used to cook food in the dining room for French service.

Recipe: A list of ingredients and a set of instructions for preparing a food or drink.

Recommended daily allowances (RDA): Average daily consumption levels of energy and selected nutrients for maintenance of health recommended for people in the United States.

Restaurant chain: A form of restaurant ownership in which individual operations are part of a multiunit group owned by a parent company, a franchise company, or independent owners.

Salamander: A piece of equipment used to brown food items or to melt cheese.

Sanitarian: Food inspector.

Sanitary: Free of harmful levels of disease-causing bacteria or contaminants.

Sanitation: The creation and maintaining of conditions favorable to good health; wholesome food is handled and prepared in such a way that it is not contaminated with disease-causing bacteria.

Sanitize: To reduce the number of harmful bacteria on a clean surface.

SAFE program: Sanitary Assessment of the Food Environment, a food safety and self-inspection system based on HACCP and developed by the National Restaurant Association.

Scientific management: Formulated by Frederick Taylor in the 1900s as one of the original forms of the study of management used to increase industrial efficiency.

Scramble system: A setup of a buffet or cafeteria line in front of which food items are spread out to allow guests to avoid crowded stations.

Seated service: A style of service in which the food is served to seated guests.

Self-service: A form of service in which guests are required to serve themselves.

Service personnel: Another name for servers or waiters.

Shadowing: A method of training in which one employee follows or "shadows" another.

Signature item: A menu item that is named after the operation or chef; usually an item that is a specialty of the operation.

Silencer cloth: The tablecloth that goes under the regular tablecloth.

Social caterers: Companies that cater social functions and parties.

Sole proprietorship: A business owned by one person.

Sous-vide: Food cooked in advance and sealed in a plastic packet with the air removed.

Specialty distributor: A foodservice supplier who carries only limited types of goods, such as fish, or poultry, or meat.

Specification: A written document about a food item that includes such things as grade, size, brand, and so on; a written purchase specification used by foodservice operations to communicate to suppliers the details of the products they wish to purchase.

Spoilage: The breakdown of the edible quality of a food product.

Stagger-cooking: Cooking some food items in batches over the shift or meal period.

Standardized recipe: A recipe that has been tested and adapted to the equipment and staff of a particular operation; standardized recipes assist in training, are an essential tool in costing and the ordering of food supplies.

Standards of uniformity: Predetermined sizes, shapes, and so on, that make it possible to use equipment produced by different manufacturers, interchange pans, racks, and so forth.

Static menu: A menu that offers the same items every day.

Steam-jacketed kettle: A piece of cooking equipment used for volume feeding. The kettle is a combination of two stainless steel bowls with a heating element and water.

Steam table: A table with a well that holds water that is heated either with electricity or gas. The wells are sized to hold special pans of food to keep them warm during serving.

Straight-line system: A cafeteria service system in which customers start at the beginning, pass all items being served, and finish at the end.

Suggestive selling: For guests arriving at a foodservice operation not knowing what they want to order, a server can influence their purchasing decision. Suggestive selling can also be effectively used to get guests to order accompaniments they might not have been aware of, such as appetizers, dessert, wine, or after-dinner drinks.

Supervisors: Generally the first step on the managerial ladder, they perform the most "hands on" work of any level of management.

Table d'hôte menu: A menu in which all courses are included at one price.

Tangible: Something that can be held or touched.

Temperature Danger Zone (TDZ): 45–140° Fahrenheit, the temperature range in which most types of bacteria flourish.

Texas Restaurant Association method: A pricing system in which menu items are separated into two categories according to how labor-intensive they are.

Theme: The organizing motif around which a restaurant's menu, decor, staff uniform, and atmosphere are developed.

Timeliness: The process of delivering food to guests in a speedy manner.

To order: Preparation of menu items in response to individual customer requests.

Top-level managers: Positions such as general manager, executive chef, or food and beverage directors. They are generally more concerned with, and responsible for, long-term planning and the goals of the operation than they are with the day-to-day operation.

Toxic: Poisonous.

Traditional foodservice system: The process in which food is served to the guest immediately after it is made.

Trailing: See *shadowing*.

Turnover: The number of people who must be hired in a year's time to keep one position filled.

Unconcerned patron: A foodservice customer who is "unconcerned" about nutrition when dining out and who orders foods that are high in fat and calories.

USDA: United States Department of Agriculture.

Vacillating patrons: Customers who change their desire for items considered to be healthy, depending on the dining occasion.

Variable costs: The costs of an operation that fluctuate with the level of sales; they include food, hourly labor, and utilities.

Walk-in refrigerator/freezer: A unit that is large enough to walk into, lined with storage shelves and cart storage locations.

Wholesomeness: A characteristic of a product that is free from disease and was processed under sanitary conditions.

Wire whip: A mixer attachment consisting of a series of heavy-duty looped wires, used to incorporate air into egg whites, cream, and so on.

Yield: The number of portions produced from a recipe.

Yield, given: The number of portions resulting from the original recipe.

Yield, new: The number of portions of a recipe needed at a specific time.

Yield test: Preparing a recipe repeatedly in order to ensure that it produces the proper number of portions.

INDEX